ISSUES IN THE INTEGRATION OF RESEARCH AND OPERATIONAL SATELLITE SYSTEMS FOR CLIMATE RESEARCH

I. Science and Design

Committee on Earth Studies
Space Studies Board
Commission on Physical Sciences, Mathematics, and Applications
National Research Council

NATIONAL ACADEMY PRESS
Washington, D.C.

NOTICE: The project that is the subject of this report was approved by the Governing Board of the National Research Council, whose members are drawn from the councils of the National Academy of Sciences, the National Academy of Engineering, and the Institute of Medicine. The members of the committee responsible for the report were chosen for their special competences and with regard for appropriate balance.

Support for this project was provided by National Aeronautics and Space Administration contract NASW-96013, and National Oceanic and Atmospheric Administration contracts 50-DGNE-5-00210 and 50-DKNA-6-90040. Any opinions, findings, conclusions, or recommendations expressed in this material are those of the authors and do not necessarily reflect the views of the sponsors.

International Standard Book Number 0-309-06985-8

Copies of this report are available free of charge from:

Space Studies Board
National Research Council
2101 Constitution Avenue, NW
Washington, DC 20418

Copyright 2000 by the National Academy of Sciences. All rights reserved.

Printed in the United States of America

THE NATIONAL ACADEMIES

National Academy of Sciences
National Academy of Engineering
Institute of Medicine
National Research Council

The **National Academy of Sciences** is a private, nonprofit, self-perpetuating society of distinguished scholars engaged in scientific and engineering research, dedicated to the furtherance of science and technology and to their use for the general welfare. Upon the authority of the charter granted to it by the Congress in 1863, the Academy has a mandate that requires it to advise the federal government on scientific and technical matters. Dr. Bruce M. Alberts is president of the National Academy of Sciences.

The **National Academy of Engineering** was established in 1964, under the charter of the National Academy of Sciences, as a parallel organization of outstanding engineers. It is autonomous in its administration and in the selection of its members, sharing with the National Academy of Sciences the responsibility for advising the federal government. The National Academy of Engineering also sponsors engineering programs aimed at meeting national needs, encourages education and research, and recognizes the superior achievements of engineers. Dr. William A. Wulf is president of the National Academy of Engineering.

The **Institute of Medicine** was established in 1970 by the National Academy of Sciences to secure the services of eminent members of appropriate professions in the examination of policy matters pertaining to the health of the public. The Institute acts under the responsibility given to the National Academy of Sciences by its congressional charter to be an adviser to the federal government and, upon its own initiative, to identify issues of medical care, research, and education. Dr. Kenneth I. Shine is president of the Institute of Medicine.

The **National Research Council** was organized by the National Academy of Sciences in 1916 to associate the broad community of science and technology with the Academy's purposes of furthering knowledge and advising the federal government. Functioning in accordance with general policies determined by the Academy, the Council has become the principal operating agency of both the National Academy of Sciences and the National Academy of Engineering in providing services to the government, the public, and the scientific and engineering communities. The Council is administered jointly by both Academies and the Institute of Medicine. Dr. Bruce M. Alberts and Dr. William A. Wulf are chairman and vice chairman, respectively, of the National Research Council.

COMMITTEE ON EARTH STUDIES

MARK R. ABBOTT, Oregon State University, *Chair*
OTIS B. BROWN, Rosenstiel School of Marine and Atmospheric Science
JOHN R. CHRISTY, University of Alabama, Huntsville
CATHERINE GAUTIER, University of California at Santa Barbara
DANIEL J. JACOB, Harvard University
CHRISTOPHER O. JUSTICE, University of Virginia
BRUCE D. MARCUS, TRW
M. PATRICK McCORMICK, Hampton University
DALLAS L. PECK, U.S. Geological Survey (retired)
R. KEITH RANEY, Johns Hopkins University Applied Physics Laboratory
DAVID T. SANDWELL, Scripps Institution of Oceanography
LAWRENCE C. SCHOLZ, West Orange, New Jersey
GRAEME L. STEPHENS, Colorado State University
FAWWAZ T. ULABY, University of Michigan
SUSAN L. USTIN, University of California at Davis
FRANK J. WENTZ, Remote Sensing Systems
EDWARD F. ZALEWSKI, University of Arizona

Staff

INA B. ALTERMAN, Senior Program Officer
ART CHARO, Senior Program Officer
CARMELA J. CHAMBERLAIN, Senior Project Assistant (to April 1999)
THERESA M. FISHER, Senior Project Assistant (from April 1999)

SPACE STUDIES BOARD

CLAUDE R. CANIZARES, Massachusetts Institute of Technology, *Chair*
MARK R. ABBOTT, Oregon State University
FRAN BAGENAL, University of Colorado
DANIEL N. BAKER, University of Colorado
ROBERT E. CLELAND, University of Washington
MARILYN L. FOGEL, Carnegie Institution of Washington
BILL GREEN, Former Member, U.S. House of Representatives
JOHN H. HOPPS, JR., Morehouse College
CHRIS J. JOHANNSEN, Purdue University
RICHARD G. KRON, University of Chicago
JONATHAN I. LUNINE, University of Arizona
ROBERTA BALSTAD MILLER, Columbia University
GARY J. OLSEN, University of Illinois at Urbana-Champaign
MARY JANE OSBORN, University of Connecticut Health Center
GEORGE A. PAULIKAS, The Aerospace Corporation
JOYCE E. PENNER, University of Michigan
THOMAS A. PRINCE, California Institute of Technology
PEDRO L. RUSTAN, JR., Ellipso, Inc.
GEORGE L. SISCOE, Boston University
EUGENE B. SKOLNIKOFF, Massachusetts Institute of Technology
MITCHELL SOGIN, Marine Biological Laboratory
NORMAN E. THAGARD, Florida State University
ALAN M. TITLE, Lockheed Martin Advanced Technology Center
RAYMOND VISKANTA, Purdue University
PETER W. VOORHEES, Northwestern University
JOHN A. WOOD, Harvard-Smithsonian Center for Astrophysics

JOSEPH K. ALEXANDER, Director

COMMISSION ON PHYSICAL SCIENCES, MATHEMATICS, AND APPLICATIONS

PETER M. BANKS, ERIM International Inc. (retired), *Co-Chair*
WILLIAM H. PRESS, Los Alamos National Laboratory, *Co-Chair*
WILLIAM F. BALLHAUS, JR., Lockheed Martin Corporation
SHIRLEY CHIANG, University of California at Davis
MARSHALL H. COHEN, California Institute of Technology
RONALD G. DOUGLAS, Texas A&M University
SAMUEL H. FULLER, Analog Devices, Inc.
MICHAEL F. GOODCHILD, University of California at Santa Barbara
MARTHA P. HAYNES, Cornell University
WESLEY T. HUNTRESS, JR., Carnegie Institution
CAROL M. JANTZEN, Westinghouse Savannah River Company
PAUL G. KAMINSKI, Technovation, Inc.
KENNETH H. KELLER, University of Minnesota
JOHN R. KREICK, Sanders, a Lockheed Martin Company (retired)
MARSHA I. LESTER, University of Pennsylvania
W. CARL LINEBERGER, University of Colorado
DUSA M. McDUFF, State University of New York at Stony Brook
JANET L. NORWOOD, Former Commissioner, U.S. Bureau of Labor Statistics
M. ELISABETH PATÉ-CORNELL, Stanford University
NICHOLAS P. SAMIOS, Brookhaven National Laboratory
ROBERT J. SPINRAD, Xerox PARC (retired)

JAMES F. HINCHMAN, Acting Executive Director

Foreword

This is the first of two reports that address the complex issue of incorporating the needs of climate research into the National Polar-orbiting Operational Environmental Satellite System (NPOESS). NPOESS, which has been driven by the imperative of reliably providing short-term weather information, is itself a union of heretofore separate civilian and military programs. It is a marriage of convenience to eliminate needless duplication and reduce cost, one that appears to be working.

The same considerations of expediency and economy motivate the present attempts to add to NPOESS the goal of climate research. The technical complexities of combining seemingly disparate requirements are accompanied by the programmatic complexities of forging further connections among three different agencies with different mandates, cultures, and congressional appropriators. Yet the stakes are very high, and each agency gains significantly by finding ways to cooperate, as do the taxpayers. Beyond cost savings, benefits include the possibility that long-term climate observations will reveal new phenomena of interest to weather forecasters, as happened with the El Niño/Southern Oscillation. Conversely, climate researchers can often make good use of operational data.

Necessity is the mother of invention, and the needs of all the parties involved in NPOESS should conspire to foster creative solutions to make this effort work. Although it has often been said that research and operational requirements are incommensurate, this report and the phase two report (*Implementation*) accentuate the degree to which they are complementary and could be made compatible. The reports provide guidelines for achieving the desired integration to the mutual benefit of all parties. Although a significant level of commitment will be needed to surmount the very real technical and programmatic impediments, the public interest would be well served by a positive outcome.

Claude R. Canizares, *Chair*
Space Studies Board

Preface

National Aeronautics and Space Administration (NASA) officials have long planned that Earth Observing System (EOS) missions would complement operational weather satellite systems, especially the Polar-Orbiting Environmental Satellites (POES) operated by the National Oceanic and Atmospheric Administration (NOAA).[1] Based on a close collaboration between NASA and NOAA, the early plans for EOS were made with the expectation that many of the EOS sensors would eventually become part of the operational observing system. However, as the plans matured, it became evident that the large facility-class instruments such as MODIS (Moderate-resolution Imaging Spectroradiometer) and AIRS (Atmospheric Infrared Sounder), desired by NASA to meet the research needs of Earth system science, would not be affordable for NOAA.

In 1996, the National Research Council's (NRC's) Committee on Earth Studies (CES) was approached by NASA to review its plans for the second series of EOS missions. Although the original plans for EOS called for repeated flights of the same sensors on all three phases to ensure data continuity,[2] NASA was then in the midst of redesigning its strategy to incorporate more flexibility so that it could take advantage of new scientific understanding as well as new technology. However, there was still an underlying need to ensure continuity of critical data sets to study climate-related processes. At the same time, NOAA and the Department of Defense had been tasked with developing a "converged" system of polar-orbiting satellites, rather than continuing to operate separate polar-orbiting meteorological satellite systems (POES and the Defense Meteorological Satellite Program—DMSP). Thus there appeared to be an opportunity to foster closer collaboration between NASA, NOAA, and DOD through the emerging National Polar-orbiting Operational Environmental Satellite System (NPOESS). Such collaboration could facilitate insertion of NASA-developed technology into the NPOESS missions as well as fulfillment of some of the EOS science requirements by the NPOESS measurements. To this end, the Integrated Program Office (IPO) for NPOESS was established to develop a joint program.

The fundamental objective of the task statement guiding this study (Appendix A) was exploration of the opportunities for a stronger relationship between the developing EOS second series (now canceled) and NPOESS to maximize the scientific opportunities for climate research. At that time, NASA's plans for EOS revolved around the continuation of 24 critical data sets. However, subsequent to definition of the original statement of

---

[1] See, for example, the chapter "EOS Program" in Ghassem Asrar and Reynold Greenstone, eds., 1995 MTPE/EOS Reference Handbook, NASA/Goddard Space Flight Center, Greenbelt, Md., 1995.

[2] EOS missions were planned to provide at least 15 years of continuous observations. After launch, each of the principal EOS spacecraft, which had an on-orbit design life of 5 years, was planned to be repeated twice.

task, NASA moved to a different approach based on key scientific questions to be developed by the Earth science community. These questions may or may not require continuity of the 24 critical data sets; NASA has engaged the Earth science community in a process to define these continuity requirements. Changes also occurred in the IPO's plans for NPOESS; in particular, the complement of sensor concepts for the satellite was fixed, thereby defining the limits of the planned observing system. The scope of the committee's potential recommendations that would be thought practical by the IPO was similarly affected, as described below.

In its letter report of May 27, 1998, "On Climate Change Research Measurements from NPOESS," CES noted that there are many scientific, technical, and programmatic issues associated with integrating the measurement responsibilities of research agencies with those of operational agencies. Using as a framework the broad area of climate research, which includes monitoring climate change as well as understanding climate processes and impacts, the committee has continued its study of these issues.

The committee uses the notion of climate observation in its broadest sense, to include monitoring climate change, understanding underlying processes, and estimating the impacts of climate change. Thus its definition extends far beyond the physical climate system; it includes biological processes as well as the linkages between the ocean, atmosphere, and land system. In this context, a satellite observing system will be required that combines elements of long-term measurements in an operational setting, systematic measurements using research satellites, and exploratory, process-oriented research missions.[3]

The committee notes that it has focused on issues relevant to climate research and acknowledges that this represents but one aspect of the broad spectrum of Earth observations for research and applications. The others also represent areas imbued with both compelling scientific merit and pressing societal urgency. Nevertheless, the committee's charge and perspective focus on climate research.

With regard to the original charge (Appendix A), the committee modified its study in response to changes in both the NASA and NPOESS strategies. Although the focus remains on the integration of research and operational missions for Earth science, the study does not consider the EOS AM-2 or PM-2 missions, which are no longer part of the NASA plan. Since IPO/NPOESS has determined its measurement suite, the study does not explicitly examine issues regarding new sensors for NPOESS. The study focuses on the additional capabilities that are required to meet climate research goals and their technical and programmatic implications, particularly for NPOESS. This phase one report also examines issues of program synchronization with regard to schedule as well as maintaining sufficient program flexibility. Lastly, the committee studied science requirements for data interoperability and continuity in the context of climate research.

To accomplish this, the committee selected for review eight representative measurement sets based on their breadth of implementation with regard to research and operational satellite missions. Some of the measurement sets have been part of the operational missions for decades, while others are just now being proposed for a transition from research to integration with the operational program. While these eight measurement sets are important for climate research, the committee is not implying that they were selected because they are the most critical measurements. Instead, these eight were reviewed to identify and highlight common issues associated with the integration of operational and research missions.

This report identifies and discusses issues related to the challenges posed by EOS and NPOESS integration; it also suggests an approach to achieve a rational balance of the available observing resources and assets that can be leveraged for climate research. The committee's forthcoming phase two report examines technical approaches to data continuity and interoperability, sensor replenishment, and the infusion of new technology.[4] The phase two report also considers issues in instrument calibration and data product validation.

---

[3]National Research Council (NRC). 1998. Overview, Global Environmental Change: Research Pathways for the Next Decade. Washington, D.C.: National Academy Press.

[4]National Research Council, Space Studies Board. 2000. Issues in the Integration of Research and Operational Satellite Systems for Climate Research: II. Implementation, forthcoming.

Acknowledgment of Reviewers

This report has been reviewed by individuals chosen for their diverse perspectives and technical expertise, in accordance with procedures approved by the National Research Council's (NRC's) Report Review Committee. The purpose of this independent review is to provide candid and critical comments that will assist the authors and the NRC in making the published report as sound as possible and to ensure that the report meets institutional standards for objectivity, evidence, and responsiveness to the study charge. The contents of the review comments and draft manuscript remain confidential to protect the integrity of the deliberative process.

We wish to thank the following individuals for their participation in the review of this report: Frederick J. Doyle, U.S. Geological Survey (retired); Charles Elachi, Jet Propulsion Laboratory; Anthony W. England, University of Michigan; John E. Estes, University of California at Santa Barbara; Richard M. Goody, Falmouth, Massachusetts; Dennis L. Hartmann, University of Washington; Jerry D. Mahlman, Geophysics Fluid Dynamics Laboratory/NOAA; John McElroy, University of Texas at Arlington; Owen M. Phillips, Johns Hopkins University; Steven Running, University of Montana; John Seinfeld, California Institute of Technology; Robert J. Serafin, National Center for Atmospheric Research; W. James Shuttleworth, University of Arizona; and Bruce A. Wielicki, NASA Langley Research Center.

Although the individuals listed above have provided many constructive comments and suggestions, responsibility for the final content of this report rests solely with the authoring committee and the NRC.

Contents

EXECUTIVE SUMMARY   1

1   INTEGRATING RESEARCH AND OPERATIONAL MISSIONS IN SUPPORT OF CLIMATE   7
    RESEARCH
    Weather and Climate, 8
    Long-term Measurements, 8
    NASA's Approach to Long-term Observations, 10
    NOAA's Approach to Long-term Observations, 11
    Joint NASA/IPO Plans, 13
    Integrating Climate Research at the Federal Level, 14
    Identifying Relevant Issues: Review of Eight Measurement Sets, 15
    References, 15

2   ATMOSPHERIC SOUNDINGS   17
    Introduction, 17
    A Brief Historical Perspective, 18
    Observing Strategies, 21
    Evolution Strategy, 22
    Challenges Ahead, 23
    References, 23

3   SEA SURFACE TEMPERATURE   25
    Introduction, 25
    Basic Science Issues, 25
    Future Directions, 28
    Observing Strategy, 29
    Calibration and Validation, 32
    Data Management, 34
    Evolution Strategy, 35
    References, 35

4 LAND COVER — 37
Introduction, 37
Basic Science Issues, 37
Future Directions, 39
Current Satellite Sampling Strategies, 40
Current Observation Systems, 40
Observing Strategies, 42
International Aspects of Land-Cover Observation, 44
What Is Needed in Addition to What Is Planned, 45
Calibration and Validation and Mission Overlap Strategies, 49
Data Processing and Management, 51
The Necessary Observation Strategy, 51
Areas for Research and Development, 52
Bibliography, 53

5 OCEAN COLOR — 57
Introduction, 57
Basic Science Issues, 57
Observing Strategy, 61
Data Products, 63
Calibration and Validation, 65
Evolution Strategy, 66
References, 67

6 SOIL MOISTURE — 68
Introduction, 68
Basic Science Issues, 69
Observing Strategy of Current and Future Satellite Sensors, 78
Calibration and Validation, 78
Evolution Strategy, 80
References, 81

7 AEROSOLS — 82
Introduction, 82
Basic Science Issues, 82
Observing Strategy, 86
Calibration and Validation Strategy, 93
Data Management, 95
Evolution Strategy, 95
Bibliography, 96

8 OZONE — 99
Introduction, 99
Basic Science Issues, 100
Observing Strategy, 105
Calibration and Validation, 106
Evolution Strategy, 107
References, 108

9	**EARTH RADIATION BUDGET**	109
	Introduction, 109	
	Radiation Budget in the Satellite Era, 111	
	Observing Strategy, 111	
	Calibration and Validation Strategies, 114	
	Opportunities, 115	
	Limitations and the Evolution Strategy, 115	
	References, 116	
10	**ISSUES, CHALLENGES, AND RECOMMENDATIONS**	117
	Common Issues, 117	
	The Challenges of Space-based Climate Research, 119	
	Recommendations, 121	
	References, 124	

APPENDIXES

A Statement of Task, 127
B Acronyms and Abbreviations, 129

Executive Summary

INTRODUCTION

Currently, the Departments of Defense (DOD) and Commerce (DOC) acquire and operate separate polar-orbiting environmental satellite systems that collect data needed for military and civil weather forecasting. The National Performance Review (NPR)[1] and subsequent Presidential Decision Directive (PDD)/NSTC-2, dated May 5, 1994, directed the DOD (Air Force) and the DOC (National Oceanic and Atmospheric Administration, NOAA) to establish a converged national weather satellite program that would meet U.S. civil and national security requirements and fulfill international obligations.[2] NASA's Earth Observing System (EOS), and potentially other NASA programs, were included in the converged program to provide new remote sensing and spacecraft technologies that could improve the operational capabilities of the converged system. The program that followed, called the National Polar-orbiting Operational Environmental Satellite System (NPOESS), combined the follow-on to the DOD's Defense Meteorological Satellite Program and the DOC's Polar-orbiting Operational Environmental Satellite (POES) program. The tri-agency Integrated Program Office (IPO) for NPOESS was subsequently established to manage the acquisition and operations of the converged satellite.

NASA officials have long envisioned developing operational versions of some of the advanced climate and weather monitoring instruments planned for EOS. In its 1995 EOS "Reshape" exercise, NASA adopted the assumption that some of the planned measurements in the second afternoon (PM) satellite series would be supplied by NPOESS. Although NASA has altered its earlier plans for the PM satellite and other follow-on missions to the first EOS series, its intent to integrate NPOESS into its Earth observation missions remains intact.

This report, the result of the first phase of a study by the Committee on Earth Studies, analyzes issues related to the integration of EOS and NPOESS, especially as they affect research and monitoring activities related to

---

[1]See DOC12: "Establish a Single Civilian Operational Environmental Polar Satellite Program," in Appendix A of From Red Tape to Results: Creating a Government that Works Better and Costs Less (National Performance Review Part I). Available on the World Wide Web at <http://www.npr.gov/library/nprrpt/annrpt/redtpe93/index.html>.

[2]"Fact Sheet: U.S. Polar-Orbiting Operational Environmental Satellite Systems and Convergence of U.S. Polar-Orbiting Operational Environmental Satellite Systems and Landsat Remote Sensing Strategy," statement by the White House Press Secretary, May 10, 1994. Available on the World Wide Web at <http://www.whitehouse.gov/WH/EOP/OSTP/NSTC/html/pdd2.html>.

Earth's climate and whether it is changing.[3] The development of high-quality, long-term satellite-based time series suitable for detection of climate change as well as for characterization of climate-related processes poses numerous challenges. In particular, achieving NASA research aims on an NPOESS satellite designed to meet the high-priority operational needs of the civil and defense communities will require agreement on program requirements, as well as coordination of instrument development activities, launch schedules, and precursor flight activities.

The study of climate processes requires a coherent, comprehensive system that carefully balances research requirements that are sometimes in conflict with operational requirements. Long-term, consistent data sets require careful calibration, reprocessing, and analysis that may not be necessary to meet the needs of short-term forecasting. Acquisition of multiple copies of a satellite sensor may be the simplest and most cost-effective means to ensure data continuity, but this strategy may preclude the insertion of new techniques to improve the observations in response to lessons learned during analysis of long data records. Such conflicts are difficult to resolve and are complicated by differences in agency cultures, charters, and financial resources.

APPROACH AND OBSERVATIONS

In performing its assessment, the committee reviewed eight variables (eight measurement areas) that it believed to be representative of the wide-ranging set of potential variables to be measured in a climate research and monitoring program. The committee adopted this strategy in part because there is no unique set of "climate variables," nor is there consensus on what might constitute a minimal set of variables to be monitored in a climate research program. The committee assessed the eight variables in terms of their value to climate science and whether the present state of measurements and their associated algorithms were adequate to produce "climate-quality" data products. Included in the committee's analysis is an assessment of the role of new technology or new measurement strategies in enhancing existing climate data products or delivering new data products of interest.

Common Issues

In its review of the eight representative climate variables the committee identified the following common issues:

- **Need for a comprehensive long-term strategy.** Systems for observing climate-related processes must be part of a comprehensive, wide-ranging, long-term strategy. Monitoring over long time periods is essential to detecting trends such as changes in sea-surface temperature and to understanding critical processes characterized by low-frequency variability. The committee notes that an observing system developed for long-term climate observations may also very well reveal unexpected phenomena, as was the case with observations of the large-scale, low-frequency El Niño/Southern Oscillation.
- **Desirability of multiple measurements of the same variable using different techniques.** Corroborating results from a variety of observing techniques increases confidence in the data; conflicting measurements suggest problems in data quality or newly emerging science questions that must be resolved.
- **Diversity of satellite observations and sampling strategies and support for ground-based networks.** While plans for NPOESS and EOS have focused primarily on polar-orbiting satellites, satellite observations from other orbits (low inclination, geostationary) have important roles in the development of a climate observing system. Differing sampling strategies will also be needed to tailor measurement requirements to instrument capabilities in a cost-effective manner.

---

[3]The committee's forthcoming phase two report, *Issues in the Integration of Research and Operational Satellite Systems for Climate Research: II. Implementation* (NRC, 2000), addresses systems engineering issues related to sensor replenishment and technology insertion, explores technical approaches to data continuity and interoperability from the standpoint of data stability, and considers issues in instrument calibration and data product validation.

Ground-based networks support and extend the space-based observations. They are critical for calibrating and validating space-based measurements; they also complement space-based measurements and often provide the high-resolution measurements in both time and space needed to carry out the process studies that elucidate the mechanisms underlying climate-related phenomena. In reviewing its notional set of eight climate variables, the committee found that more attention to development of ground-based networks was warranted.

- **Preserving the quality of data acquired in a series of measurements.** A particular challenge in the design of a climate observing systems is how to preserve data quality and facilitate valid comparisons of observations that extend over a series of spacecraft. With the regular insertion of new technology driven by interest in reducing costs and/or improving performance also comes the need to separate the effects of changes in the Earth system from effects ascribable to changes and gaps in the observing system. Effective, ongoing programs of sensor calibration and validation, sensor characterization, data continuity, and strategies for ensuring overlap across successive sensors are thus essential. Data systems should be designed to meet the need for periodic reprocessing of the entire data set.
- **The role of data analysis and reprocessing.** An active, continuous program of data analysis and reprocessing adds value to existing data sets and enables the development of new algorithms and new data products.
- **Technology development and improved measurement capabilities.** New sensors are needed to reduce costs and to improve existing measurement capabilities. In addition, some climate-related variables, for example, soil moisture, cannot be measured adequately with existing capabilities. Moreover, it is not clear that all critical climate-related variables have even been identified. With improved coordination with NOAA and the IPO for NPOESS, NASA technology development efforts would better address these issues and help provide increased capabilities for the operational meteorological system.

Carrying Out Climate Research from Space-Based Platforms

Operational agencies generally respond to short-term demands for data products; research agencies are also under increasing pressure to respond to short-term demands for technology development and science missions that can be accomplished in a few years. As a result, political and programmatic pressures for short-term returns (both in terms of science and protection of life and property) have resulted in an operational agency focus on the acute problems of storms, earthquakes, and other severe events—even though there is growing evidence that the long-term trends associated with climate will have significant economic and social impacts. Addressing the issues associated with climate will require a long-term focus and a commitment to maintain long-term, high-quality observing systems.

Climate research and monitoring require a blend of short-term, focused measurements as well as systematic, long-term measurements. While the generally shorter-term and more detailed studies that characterize process studies might appear to be in opposition to a long-term program of systematic measurements, the committee emphasizes that climate-related processes are often revealed only through the study of data from long-term systematic measurements. Achieving an appropriate balance across agencies between short-term and long-term activities related to climate research, such as a balance between process studies and monitoring activities, has proved difficult. Recent NRC studies have recommended that the Executive Branch establish an office to develop and manage a climate observing strategy.[4]

NPOESS and Climate Research

The 1994 Presidential Directive to converge DOD and DOC meteorological programs initiated a lengthy process among Air Force and NOAA operational and research users to produce a detailed list of measurement requirements. The culmination of this effort was the *Integrated Operational Requirements Document* (IORD-1)

---

[4]See, for example, NRC (1998, 1999b).

that was formally endorsed by NOAA, DOD, and NASA.[5] The IORD-1 consists of 61 environmental data records (EDRs) deemed necessary to the success of NPOESS. The EDRs are distributed among six categories: atmospheric parameters, cloud parameters, Earth radiation budget parameters, land parameters, ocean and water parameters, and space environmental parameters.

The EDRs developed in the IORD-1 describe a well-defined, detailed set of measurements that have demonstrable value in the primary NPOESS mission of short-term weather forecasting. Climate research and modeling, however, require assimilation and analysis of a much broader set of measurements that may also be characterized by different time and space scales. Instrument stability is a key consideration in the analysis of whether climate variables are changing, yet it is undefined for many of the EDRs. Further, the IORD-1 does not set requirements on the stability or longevity of the stipulated measurements.

Despite these problems, the committee believes that NPOESS offers a unique opportunity to establish a satellite-based observing system for climate research and monitoring. Although the NPOESS and NASA EOS missions as currently planned may not be optimum for climate research, many of the critical components are already in place. These include an initial commitment to data stability on the part of the NPOESS IPO, an active program of data analysis and data product validation by NASA's Earth Science Enterprise (ESE), and an active plan for NASA and NOAA collaborative missions such as the NPOESS Preparatory Project. The committee is concerned, however, that budget pressures, shifting programmatic interests, and a lack of overall vision and leadership may continue to inhibit the establishment of a coherent Earth observing system for climate research and monitoring.[6]

Challenges in the Integration of NASA/ESE and NOAA/NPOESS Programs

- **Division of responsibility in the integration of research and operational missions.** Climate research and monitoring raise issues that transcend the capabilities of any single federal agency. Yet, in the committee's view, no effective structure is currently in place in the federal government that can address such multiagency issues as the balance between satellite and ground-based observations, long-term and exploratory missions, and research and operational needs. The committee concurs with recent NRC reports that have expressed concern over the lack of overall authority and accountability, the division of responsibility, and the lack of progress in achieving a long-term climate observing system.[7] The challenges in integrating ESE research satellite missions and NPOESS operational satellite missions underscore these critical issues.

- **Adequacy of NPOESS environmental data requirements for climate research.** The EDR process established by the IPO supports the primary operational goals of DOD and NOAA but was not intended to yield instrument specifications that meet climate research requirements. For example, many climate research studies require access to unprocessed sensor-level data, whereas the EDR approach focuses on the final data products. In many cases, the current EDRs are not completely specified, and in some, they are not adequate for climate research. A particular issue is the absence of measurement stability and longevity specifications for many of the EDRs.

- **Ensuring the long-term (systematic) record begun by EOS.** NASA's ESE plans that certain measurements begun on EOS satellites will be integrated later into the NPOESS program. However, given the budgetary and programmatic uncertainties that have historically characterized the EOS program, there can be no assurance that this integration will be successful. Further, the committee notes that while long-term observations are essential for climate studies, NASA's new EOS plan focuses on short-term (3- to 5-year) missions. For NASA to be able to pursue a science-based strategy that leverages NPOESS capabilities where possible, the agency will probably also have to fly complementary missions and collect specialized data sets.

---

[5]An updated IORD and other documentation related to the NPOESS program are available online at <http://npoesslib.ipo.noaa.gov/ElectLib.htm>.

[6]An additional set of issues relates to the development of suitable long-term climate data archive, the subject of another study by the committee, *Ensuring the Climate Record from the NPP and NPOESS Meteorological Satellites*, currently in press.

[7]See, for example, NRC (1998, 1999a,b).

Satellite observing systems are developed for a range of objectives that sometimes conflict, leading to the need for a framework to evaluate trade-offs and to manage risk. The NPOESS Preparatory Project (NPP) under consideration by NASA and the IPO is an encouraging step toward addressing the need to maintain continuity of critical data sets between the end of the EOS platforms and the launch of the first NPOESS platforms.

- **Development of sustainable instrumentation.** Sensors developed for NASA ESE research missions are generally intended to make ambitious state-of-the-art measurements. They are typically relatively complex and often are developed in small numbers, or even as one of a kind. In contrast, sensors for operational weather forecasting missions are generally less expensive to build and operate and are designed with reliability as a key requirement. Repeat flights of identical sensors are typical in NOAA operational meteorology programs. Developing instruments appropriate for both research- and operational-type missions that can be sustained over the longer periods characteristic of a climate research program will be a particular challenge as EOS and NPOESS satellites are integrated.
- **Prioritizing and establishing an observing strategy.** The climate research community has not yet prioritized critical data sets or developed an overall national observing strategy, including algorithm development, calibration and validation, ground observations, and new technology. Climate research priorities should reflect scientific need, while recognizing technological, fiscal, and programmatic constraints. Other important aspects of such a strategy will be periodic evaluation and readjustment of specific mechanisms for transferring data sets from research to operations. Articulation of a long-term context, spanning as much as a century or more, will be paramount in developing a credible climate observing policy.

RECOMMENDATIONS

The following recommendations are directed to the climate research community, NASA's Earth Science Enterprise, and the NPOESS Integrated Program Office. They derive from consideration of the common issues associated with the space-based measurement of climate variables and committee concerns related to the conduct of climate research.

Recommendation 1.

Climate research and monitoring capabilities should be balanced with the requirements for operational weather observation and forecasting within an overall U.S. strategy for future satellite observing systems. The committee proposes the following specific actions to achieve this recommendation:

- *The Executive Branch should establish a panel within the federal government that will assess the U.S. remote sensing programs and their ability to meet the science and policy needs for climate research and monitoring and the requirements for operational weather observation and forecasting.*
  —*The panel should be convened under the auspices of the National Science and Technology Council and draw upon input from agency representatives, climate researchers, and operational users.*
  —*The panel should convene a series of open workshops with broad participation by the remote sensing and climate research communities, and by operational users, to begin the development of a national climate observing strategy that would leverage existing satellite-based and ground-based components.*

Recommendation 2.

The Integrated Program Office for NPOESS should give increased consideration to the use of NPOESS for climate research and monitoring. The committee proposes the following specific actions to achieve this recommendation:

- *The IPO should consider the climate research and monitoring capabilities of NPOESS along with other NPOESS requirements.*
- *For those NPOESS measurements that are deemed to be critical for climate research and monitoring, the IPO should establish a science oversight team with specific responsibilities for each associated sensor suite.*
- *The IPO should begin to establish plans for sensor calibration and data product validation as well as for data processing and delivery that consider the needs for climate research.*

Recommendation 3.

The NASA Earth Science Enterprise should continue to play an active role in the acquisition and analysis of systematic measurements for climate research as well as in the provision of new technology for NPOESS. The committee proposes the following specific actions to achieve this recommendation:

- *NASA/ESE should develop specific technology programs aimed at the development of sustainable instrumentation for NPOESS.*
- *NASA/ESE should ensure that systematic measurements that are integrated into operational systems continue to meet science requirements.*
- *NASA/ESE should continue satellite missions for many measurements that are critical for climate research and monitoring.*

Recommendation 4.

Joint research and operational opportunities such as the NPOESS Preparatory Project should become a permanent part of the U.S. Earth observing remote sensing strategy. The committee proposes the following specific actions to achieve this recommendation:

- *The NPP concept should be made a permanent part of the U.S. climate observing strategy as a joint NASA-IPO activity.*
- *Some space should be reserved on the NPOESS platforms for research sensors and technology demonstrations as well as to provide adequate data downlink and ground segment capability.*
- *NPP and NPOESS resources should be developed and allocated with the full participation of the Earth science community.*

REFERENCES

National Research Council (NRC). 1998. Overview, Global Environmental Change: Research Pathways for the Next Decade. Washington, D.C.: National Academy Press.

National Research Council (NRC). 1999a. The Adequacy of Climate Observing Systems. Washington, D.C.: National Academy Press.

National Research Council (NRC), Space Studies Board. 1999b. "Assessment of NASA's Plans for Post-2002 Earth Observing Missions," short report to Dr. Ghassem Asrar, NASA's Associate Administrator for Earth Science, April 8.

National Research Council, Space Studies Board. 2000. Issues in the Integration of Research and Operational Satellite Systems for Climate Research: II. Implementation. Washington, D.C.: National Academy Press, forthcoming.

1

Integrating Research and Operational Missions in Support of Climate Research

That Earth's climate has changed and that it will continue to do so is well appreciated. Nevertheless, there are persistent questions regarding present climate trends on a decadal or centenary time scale and the appropriate policies for responding to climate change. Within this framework, there is the need both to determine the processes controlling climate change and variability and to monitor climate change.

There is occasionally the perception that *process studies* require observing systems that are different from systems for *monitoring*. This is not always the case, especially with regard to studies involving climate variability. For example, study of the processes underlying the El Niño/Southern Oscillation (ENSO) requires a variety of data types comprising consistent observations over many years. Thus, the requirements for process studies and monitoring tend to merge as the time scale of the phenomena of interest increases.

Both operational and research satellite systems have played an important role in the development of our understanding of Earth processes (NRC, 1995). Although the primary focus of the operational systems is on short-term weather prediction and the protection of life and property, they have also played a vital role in study of the Earth. The operational nature of these systems has ensured that the data record is nearly complete, spanning more than 30 years in certain cases. Many of the variables important for weather forecasting are also critical for understanding climate-related processes as well as climate monitoring. Sea surface temperature (SST) is a notable example of such a parameter. It is frequently used as a diagnostic variable in global circulation models, as well as input in short-term weather forecasts. It is also used in commercial applications, such as identifying prime fishing areas. In contrast, research observing systems are relatively short-lived (less than 5 years) and focus on a specific scientific or technical issue. However, there are many examples where research missions survive for long time periods or involve repeated flights of the same research sensor. For example, copies of the Total Ozone Mapping Sensor (TOMS) have been flown for nearly 20 years, providing a valuable record of long-term changes in stratospheric ozone.

Monitoring climate change has stringent requirements. Depending on the time scale of interest and the nature of the particular process, the signal may be small relative to other sources of variability. For example, it has been projected that global SST will increase 0.25 °C per decade in response to increasing atmospheric concentrations of CO_2. However, ENSO events will dominate SST variability on interannual time scales. This implies that the SST record will have to be compiled over many years and with high precision and accuracy to detect this projected response,

WEATHER AND CLIMATE

The requirements for a program of climate research are often perceived to be in conflict with what is required for weather services: climate-related studies require long-term consistency whereas weather services (e.g., forecasts, severe weather warnings) require rapid delivery of data products. However, climate can be viewed as the long-term statistics of weather. This linkage may be exploited to meet both sets of requirements. Thus variables that are critical for weather forecasting, such as atmospheric profiles of temperature and humidity, are also important for climate research and modeling.

One of the fundamental differences is the time scale over which both the forecasts and the observations must be made. For example, simple weather forecasts based on persistence (i.e., tomorrow's weather will be the same as today's) work well over short time scales. On longer time scales, more complicated models and a richer observation suite are required. For example, ocean processes (such as ocean circulation) and terrestrial processes (such as evapotranspiration) need to be included to produce forecasts with sufficient capability or to understand the critical processes. It is possible to generalize by noting that as the time scale of interest increases, more processes and more complicated interactions become important. This effect leads to the occasional surprises noted in the overview of the *Pathways* report (NRC, 1998a)—the unexpected processes or linkages that appear in studies of climate change. Box 1.1 elaborates further on the distinctions between weather and climate.

Nonlinear processes and the increasing number of interacting processes make it impossible to define a priori all of the types and scales of observations that need to be made. Accordingly, there should be balance between the focused research missions where the scientific underpinnings are well known and the wide-open, broadly based observations of some operational missions. The *Pathways* report overview (NRC, 1998a) discusses a scientific framework to support an observing strategy. This framework builds on the first decade of the U.S. Global Change Research Program (USGCRP) and identifies several areas of science and observations where a renewed focus and a rebalancing of priorities are required.

A noteworthy inference that can be drawn from the *Pathways* report is that establishing a robust understanding of the linkages between large-scale global processes and smaller-scale regional processes is an enormous challenge. For example, changes in ecosystem structure may be linked to changes in the patterns of climate variability, which in turn have feedbacks on the climate system. Moreover, public policy will respond to such regional-scale impacts rather than to broad-scale global change in mean Earth system properties. The Earth Observing System (EOS) and National Polar-orbiting Operational Environmental Satellite System (NPOESS) missions need to accommodate this scientific framework and balance the often conflicting primary missions of operational and research systems, the needs for continuity and innovation, and the needs for process studies and long-term monitoring.

LONG-TERM MEASUREMENTS

The characteristic scales of climate variability demand long time series in order to determine the critical processes as well as to separate natural variability from anthropogenic influences. Unlike weather forecasting, the interval between stimulus and response can be several years to centuries. With a high level of background variability, subtle changes in Earth's climate system can be difficult to detect. This problem is further complicated by the changes in instrument technology or sampling strategies that may occur during the period of observations.

The task of assembling a record of total solar irradiance (Willson and Hudson, 1991) illustrates the challenges facing the development of long-term consistent time series. Developing this record required a rigorous calibration and sensor characterization program and an observation approach that ensured sufficient temporal overlap (as well as sensor validation) to achieve accurate cross-calibration between successive sensors. Another notable example is the record of upper-troposphere temperature (NRC, 2000).

In the science community, long-term data sets are sometimes perceived as being the result of unchanging data collection activities that are not at the forefront of innovative research. It is difficult to base a scientific career on such an activity. Nevertheless, the atmospheric CO_2 record started by C.D. Keeling at Mauna Loa shows the importance of such long-term records and how the scientific value of such time series increases as the record

> **Box 1.1**
> **Distinguishing Weather and Climate**
>
> "Weather," the current condition of the atmosphere, is usually described in terms of temperature, cloud cover, wind, and precipitation. Operational weather forecasts attempt to predict the evolution of these variables over the next few hours or days at specific locations to answer questions such as, Will it rain today or tomorrow? or Am I threatened by severe weather?
>
> "Climate" research focuses on long time scales and addresses questions such as, How does today's weather compare with that of a decade, a century, or a millennium ago? Are there long-term trends in regional and global variables? If so, why? Will there be another Ice Age or human-induced global warming? The signal of these types of fluctuations can be extremely small compared with daily weather changes. For example, the "climate" variation of globally averaged decadal temperatures over the past 500 years has been only about 1 °C. It is not uncommon, however, to observe day-to-day variations in local "weather" that produce temperature changes of many times that amount.
>
> Weather forecasting and climate research, therefore, place different demands on data and consequently necessitate different strategies for making and utilizing observations. The key factors distinguishing the strategies are time scale and precision. Operational weather observations require near-real-time access to data for rapid processing so that the current state of the atmosphere can be adequately characterized in terms of physical variables in order that computer models may provide timely projections. The interval from taking an observation to processing it is often less than 1 hour. Climate researchers, though, have the luxury of time to sift through the observations (if the data have been archived and are accessible) in order to assess their precision and utility.
>
> Virtually all observations needed for operational weather forecasts, if properly calibrated, are valuable for climate research, but several climate-sensitive parameters have little bearing on the weather forecast for the next week or so. Such variables as ocean topography and salinity, ice-cap thickness, volcanic activity, stratospheric temperatures, atmospheric chemical composition, and ground cover, for example, are generally treated as constants for the operational forecast period. However, small variations in these components are critical for understanding longer-term climate issues. Monitoring climate variables requires a permanent commitment to systematic observations, some of which have little immediate value for the task of weather forecasting.
>
> In addition, because operational weather forecasting is generally the mission of a national or multinational institution, conventional observations (e.g., of rainfall or surface temperature) tend to be clustered in the forecast area and often are not made systematically across institutions. Climate fluctuations occur on a global scale, and characterizing them requires uniformly distributed and systematically observed data collected without regard for national boundaries or human schedules.

length increases (Keeling et al., 1996). Moreover, the Mauna Loa record also is an excellent case study of the programmatic difficulty in maintaining such time series. The level of personal and political commitment needs to be high, and short-term funding strategies and traditional peer review often work in opposition.

Thus the science community often considers such long time series to be the purview of the operational agencies where long-term funding can be sustained. There may be negative consequences to such a strategy. First, it may separate data collection from active research, so that the process degrades to passive monitoring. Second, operational agencies sometimes have constrained budget flexibility, and with good reason they are reluctant to assume a continuing mandate for data collection without sufficient resources. The more that long-term time series are entitlements in an agency budget, the less flexibility an agency may have to pursue new activities.

NASA'S APPROACH TO LONG-TERM OBSERVATIONS

The original plans for EOS included three nearly identical groups of satellites, with each group lasting 5 years. The resulting 15-year data set would form the basis for climate research and modeling. Implicitly, it was assumed that a subset of these observations and their associated requirements would find their way into the operational observing systems and continue indefinitely. In some cases, such a research/operational partnership was developed explicitly; for example, the afternoon platform, PM-1, was assumed to be of particular interest for both operational weather and climate research applications. The Atmospheric Infrared Sounder (AIRS) was one example of a NASA sensor that was expected to find a home in the NOAA POES.

As the plans for EOS shifted in response to changing budget and scientific pressures, the idea of repeated flights of similar platforms and sensors was dropped. Instead, NASA focused on continuity of 24 key measurements (Table 1.1), which could perhaps be achieved by a variety of sensors during the 15-year EOS program. NASA proposed that the original EOS measurements could be divided into two categories: *process* measurements that would last only for a limited time period and *monitoring* measurements that would need to be maintained throughout the life of the EOS program. The process for dividing the EOS observation set into these two categories was begun in 1995, but it was never brought to fruition.

In briefings to the committee, NASA officials described a new process for defining the second series of EOS missions. The Earth science community will propose science-driven mission concepts. NASA officials expressed their intention to base mission design more directly on science questions than may have been the case previously. The missions would begin operation in 2004, 5 years after the launch of the first EOS platform, Terra (formerly known as AM-1). It is expected that the new missions will be considerably cheaper than the first series.

TABLE 1.1 The 24 Measurements Planned for EOS

Discipline	Measurement
Atmosphere	Cloud properties
	Radiative energy fluxes
	Precipitation
	Tropospheric chemistry
	Stratospheric chemistry
	Aerosol properties
	Atmospheric temperature
	Atmospheric humidity
	Lightning
Land	Land cover and land use change
	Vegetation dynamics
	Surface temperature
	Fire occurrence
	Volcanic effects
	Surface wetness
Ocean	Surface temperature
	Phytoplankton and dissolved organic matter
	Surface wind fields
	Ocean surface topography
Cryosphere	Land ice change
	Sea ice
	Snow cover
Solar Radiation	Total solar irradiance
	Ultraviolet spectral irradiance

Current NASA plans include continuation of a subset of what NASA has designated "systematic measurements" as well as a transition of other measurements to operational programs such as NPOESS. A key component of this strategy is the NPOESS Preparatory Project (NPP), which is tentatively scheduled for launch in 2006. The NPP will carry a subset of the NPOESS sensors (the Visible/Infrared Imager and Radiometer Suite (VIIRS), Conical Scanning Microwave Imager/Sounder (CMIS), Advanced Technology Microwave Sounder (ATMS), and Cross-Track IR Sounder (CrIS)); NASA has been working with the IPO to improve these sensors to meet some of the EOS science requirements (e.g., to incorporate some of the MODIS capabilities for visible imaging in VIIRS). Continuity of other missions (such as high-resolution land imaging) in the post-EOS era is more problematic.

NASA has moved from its rigid plan of flight copies of EOS hardware to a program that is much more flexible and in some sense less predictable. Unlike the days of the Operational Satellite Improvement Program in which NASA flight-tested hardware for the National Oceanic and Atmospheric Administration (NOAA) weather satellites (NRC, 1995), there is no structured program in NASA to develop sensors for use in NPOESS. This does not mean that such transfers cannot happen; a conscious effort on the part of the two programs could facilitate such collaboration. In fact, NASA has stated that it will not develop any sensors for the operational agencies unless there is a clear commitment to continued flight of such a sensor after its initial demonstration.

NOAA'S APPROACH TO LONG-TERM OBSERVATIONS

The 1994 Presidential Decision Directive directing the Department of Defense and the Department of Commerce to develop a converged Defense Meteorological Satellite Program/Polar-orbiting Operational Environmental Satellites program[1] initiated a process to identify joint agency requirements for the combined system. This process involved operational and research users, both internal and external to the two programs. Not surprisingly, agreement on joint agency requirements was difficult as the DMSP and POES programs serve distinct user communities. Eventually, the agencies codified their agreement on the requirements for NPOESS in the *Integrated Operational Requirements Document* (IORD-1; IPO NPOESS, 1996). The IORD-1 consists of 61 environmental data records (EDRs) that were deemed necessary to the success of NPOESS (see Table 1.2). Of these, 6 were defined by the DOD as mission-critical.

Unlike the more flexible research requirements that characterize NASA missions, the EDR process relied on a well-defined set of measurements that had demonstrable value for the primary mission of NPOESS. For example, weather forecasting models are relatively mature, and it is fairly straightforward to quantify the improvements in model predictive skill given the availability of a particular data set. The user community for NPOESS data is well defined, and specific EDRs were often developed to meet their application needs. On the other hand, climate modeling is far more complex and less advanced, so it is difficult to quantify the effects on predictive skill. Observations for climate research tend to be broad in scope, with the expectation that new insights will be gained based on the availability of long-term, well-calibrated data. Moreover, it is unrealistic to expect that climate science requirements can be met simply through a better definition of measurement requirements for NPOESS. Climate research involves far too many processes at a wide range of time and space scales.

For each EDR, both a "threshold" and an "objective" were defined by the IPO. The threshold refers to the minimum set of standards that must be met for the measurement to be a success. The objective refers to the desired standards. For many climate change applications, instrument stability is a key standard, yet it is undefined for many of the EDRs. Typically, each EDR defines measurement quality and sampling characteristics without specifying the measurement technology.

In March 1996, NOAA sponsored a workshop that brought together a broad panel of climate research scientists to evaluate the applicability of the NPOESS EDRs for climate studies (NOAA, 1997). In general, the NPOESS measurements meet some climate research requirements in terms of accuracy. However, many climate

---

[1]"Fact Sheet: U.S. Polar-Orbiting Operational Environmental Satellite Systems and Convergence of U.S. Polar-Orbiting Operational Environmental Satellite Systems and Landsat Remote Sensing Strategy," statement by the White House Press Secretary, May 10, 1994. Available on the World Wide Web at <http://www.whitehouse.gov/WH/EOP/OSTP/NSTC/html/pdd2.html>.

TABLE 1.2 Environmental Data Records for NPOESS

Discipline	Measurement
Key parameters (essential baseline measurements that must be provided by NPOESS)	Atmospheric vertical moisture profiles Atmospheric vertical temperature profiles Imagery Sea surface temperature Sea surface winds Soil moisture
Atmospheric parameters	Aerosol optical thickness Aerosol particle size Ozone total column/profile Precipitable water Precipitation (type and rate) Pressure (surface and profile) Suspended matter Total water content
Cloud parameters	Cloud base height Cloud cover and layers Cloud effective particle size Cloud ice water path Cloud liquid water Cloud optical depth and transmittance Cloud top height Cloud top pressure Cloud top temperature
Earth radiation budget parameters	Surface albedo Downward longwave radiation at the surface Insolation Net shortwave radiation at the top of the atmosphere Solar radiance Total longwave radiation at the top of the atmosphere
Land parameters	Land surface temperature Normalized difference vegetation index Snow cover and depth Vegetation and surface type
Ocean and water parameters	Currents Freshwater ice motion Ice surface temperature Littoral sediment transport Net heat flux Ocean color and chlorophyll Ocean wave characteristics Sea ice age and motion Sea surface height and topography Surface wind stress Turbidity

continued

TABLE 1.2 Continued

Discipline	Measurement
Space environmental parameters	Auroral boundary
	Auroral energy deposition
	Auroral imagery
	Electric field
	Electron density profiles and ionospheric specification
	Geomagnetic field
	In situ ion drift velocity
	In situ plasma density
	In situ plasma fluctuations
	In situ plasma temperature
	Ionospheric scintillation
	Neutral density profile/neutral atmosphere
	Radiation belt and low energy solar particles
	Solar and galactic cosmic ray particles
	Solar extreme ultraviolet flux
	Supra-thermal through auroral energy
	Upper atmospheric airglow

research objectives require that NPOESS sensors meet the EDR objective requirements, not simply the EDR threshold. In addition, many research requirements depend on the details of the technical implementation, which are not captured in the EDR. For example, measuring sea surface topography requires precise knowledge of satellite orbit and tides, which are not discussed in the IORD. Once a contractor and design have been selected, important information on the proposed instrument characteristics and associated product algorithms will become available to the climate research community for evaluation and peer review. However, unless a flexible contract is negotiated that will allow changes to be made to the design and the algorithms, considerable costs could be incurred by accommodating the climate community needs.

The IPO has specified stability requirements for many variables in the IORD-1. These variables include cloud effective particle size, cloud-top pressure, cloud-ice-water path, cloud optical depth, cloud-top height, cloud-top temperature, total column ozone and ozone profile, aerosol particle size, aerosol optical thickness, albedo, and normalized difference vegetation index (NDVI). Although these stability requirements are not complete and have not been reviewed by the climate research community to ensure their adequacy, they do represent an important shift in the direction of operational satellite systems. The IORD-1 refers to NOAA's "climate monitoring mission" and notes that the U.S. government requirements include "seasonal and interannual climate forecasts; . . . decadal-scale monitoring of climate variability; . . . [and] assessment of long-term global environmental change" as part of NPOESS (IPO NPOESS, 1996). However, there is far more to climate monitoring and research than simply collecting data (NRC, 1999a,b); the "culture" required for climate observation is fundamentally different from the one that obtains for short-term forecasts.

The IPO has awarded two contracts for each of the candidate sensors. There have been final selections of the winning contractors in 1999 and 2000. The Ozone Mapping and Profiling Suite (OMPS) contractor was selected in early 1999. The "need date," which is the date by which the first NPOESS platform must be ready to launch, is in 2003. However, the scheduled launch date is not until 2009.

JOINT NASA/IPO PLANS

NASA and IPO have begun plans for the NPOESS Preparatory Project, which would launch in 2005. The mission would support early flights of the Visible/Infrared Imager Radiometer Suite (VIIRS), the Cross-Track Infrared Sounder (CrIS), and the Advanced Technology Microwave Sounder (ATMS). Other small sensors also may be flown. NPP will support early testing and evaluation of critical instruments and algorithms for NPOESS. Research requirements for selected NASA data sets have also been included in the NPP sensor requirements. The NPP mission may also demonstrate advanced technology options developed by NASA. Thus, NPP provides the opportunity to blend science and operational requirements while bridging between the first EOS series and NPOESS.

INTEGRATING CLIMATE RESEARCH AT THE FEDERAL LEVEL

At the federal level, the USGCRP was created as a "virtual agency"[2] to coordinate the activities of NASA, NOAA, the National Science Foundation, DOE, and other agencies concerned with monitoring, predicting, or responding to potential changes in Earth's global environment. The cross-agency coordination of the USGCRP is conducted under the auspices of a subcommittee that reports to the White House-level National Science and Technology Council. Since its inception, issues related to global climate change have been identified as the highest priority for research coordinated through the USGCRP. However, recent NRC panels have faulted the USGCRP and/or agencies participating in the USGCRP regarding a variety of issues related to development of the required long-term observing and data management systems.[3] A recurrent theme in these reports is the enormous technical and programmatic difficulties in assembling a climate observing system based on research and operational assets.

NASA, which has a particularly important role in the USGCRP, has announced its intention to devote greater resources to the study of "climate forcing, climate response, and the processes connecting the two."[4] However, NASA also acknowledges the necessity of exploring new arrangements with its USGCRP partners to develop a credible observing system suitable for climate research. Responding on behalf of NASA to the findings of the "Post-2002" report (NRC, 1999b), an official stated,

> The NRC Task Force noted appropriately that no single federal agency or administration is currently mandated to develop and operate for the appropriate period of time in the future, the full range of observations that are needed to understand and predict the behavior of the global Earth environment. NASA and NOAA simultaneously took the initiative to call the attention of the Executive Branch to this problem and are currently engaged in consultations with the Office of Science and Technology Policy to lay the foundations of a federal policy on this matter.[5]

The *Pathways* report overview (NRC, 1998a), based on the NRC Committee on Global Change Research assessment of the USGCRP, emphasized the critical nature of high-quality, long-term observations of the Earth system from both a scientific and public policy perspective. NPOESS and EOS are critical elements of this strategy, but the need to observe decadal and longer-term changes raises basic issues of observing system design and management.

---

[2]"Virtual agency" refers to the USGCRP interagency body. See p. ii in USGCRP (1997).

[3]For example, the *Pathways* report (NRC, 1998a) noted that "correctly transferring . . . key aspects of the observing program for USGCRP to operational programs will be very difficult." The Climate Research Committee (NRC, 1999a) in its report *The Adequacy of Climate Observing Systems* stated that "there has been a lack of progress by the federal agencies responsible for climate observing systems, individually and collectively, toward developing and maintaining a credible integrated climate observing system." The "Post-2002" report (NRC, 1999b) determined that "ensuring continuity of operational data, evaluating the readiness of a given 'research' data series to move to an operational status, and managing the 'research-to-operations' transition of data are problems that will require scientific community involvement and NASA leadership among the USGCRP agencies." In its report *The Atmospheric Sciences: Entering the Twenty-First Century*, the Board on Atmospheric Science and Climate (NRC, 1998b) noted that ". . . a comprehensive climate research program that serves societal needs is clearly within our grasp."

[4]"Understanding Our Home Planet: NASA's Role in Studying Global Climate Change," remarks of NASA Administrator Daniel S. Goldin to the 80th Annual Meeting of the American Meteorological Society, January 9, 2000.

[5]"NAS/NRC Review of NASA's Plans for Post-2002 Earth Observation Missions," briefing by NASA to Board on Atmospheric Sciences and Climate, Woods Hole, MA, June 29, 1999. See also letter from Mr. Daniel S. Goldin, Administrator of NASA, to Dr. Neal Lane, Director, White House Office of Science and Technology Policy, February 1, 1999.

IDENTIFYING RELEVANT ISSUES: REVIEW OF EIGHT MEASUREMENT SETS

Issues related to the development of a coherent national strategy for climate observations are addressed in this report in the context of a subset of measurements of demonstrable importance to climate research. With the emergence of NPOESS, the increased interest in NASA collaboration with operational agencies, and recognition of the importance of observations for climate research and policy, there is an opportunity to build the space-based component of a climate observing system. In collaboration with NASA, these requirements can be extended and refined so that it will be possible to begin to assemble credible time series of climate-related variables.

In Chapters 2 through 9, the committee reviews eight Earth science data sets, discussing each in terms of its value for climate research and the associated requirements. The committee reviews the current status of these measurements and their associated algorithms, and also explores what the status of the requirements might be in 20 years (roughly at the end of the planned NPOESS program), when new technologies or new sampling strategies may have enhanced current data sets or enabled the delivery of new data products. *The committee emphasizes that these eight measurement sets, although critical for climate research, are not necessarily the most important. Instead, the committee reviews these measurement sets to elucidate the scientific and programmatic issues associated with the integration of research and operational systems.*

The eight variables were selected because they span a broad range of science issues and also because of the range of the strategies for their implementation. The first three (atmospheric sounding, sea surface temperature, and land cover) have been part of the operational POES program for decades. Each data set has been used in climate and global change research as well as operational programs. The second three (ocean color, soil moisture, and atmospheric aerosols) have been part of the NASA research missions and are proposed for inclusion in NPOESS. These variables have not been measured as part of a long time series but on single missions instead (e.g., ocean color) or else the data product is still in development (e.g., aerosols). The final two variables (stratospheric ozone and Earth radiation budget) have been part of a long series of research missions (e.g., TOMS), and although there are counterparts in the operational missions, the Earth science community has focused primarily on the research missions. For each of the three sets, there are planned improvements in EOS.

For each variable, the committee reviews current NASA and NPOESS plans for data collection and briefly discusses the primary sensors and their expected performance. (For some variables, international or commercial data sets may be relevant as well.) It also evaluates observing strategy in terms of data continuity and the types of data products that will be developed, and it compares strategies for calibration and validation[6] with the plans currently laid out for the relevant NASA and NPOESS missions.

Reconciling the sometimes conflicting requirements of operations and research is a difficult task, and attempts to develop a coherent, comprehensive observing strategy often have relied on ad hoc solutions. With changes in schedules, in program structure, and in fiscal resources, it has been difficult to maintain effective coordination for a sufficient period of time. For each measurement set examined in the following chapters, and in its summation in Chapter 10, the committee highlights those areas where investments or changes in management structure may help us to realize the potential for an integrated observing system for climate research.

REFERENCES

Keeling, C.D., J.F.S. Chin, and T.P. Whorf. 1996. Increased activity of northern vegetation inferred from atmospheric CO_2 measurements. Nature 382: 146-149.

Integrated Program Office (IPO), National Polar-orbiting Operational Environmental Satellite System (NPOESS). 1996. Integrated Operational Requirements Document (IORD) I. Joint Agency Requirements Group Administrators. 61 pp. + figures.

National Oceanic and Atmospheric Administration (NOAA). 1997. Climate Measurement Requirements for the National Polar-orbiting Operational Environmental Satellite System (NPOESS), Workshop Report, Herbert Jacobowitz (ed.), Office of Research and Applications, NESDIS-NOAA, Washington, D.C.

---

[6]Calibration is the process of quantitatively defining the system responses to known, controlled signal inputs, and validation is the process of assessing by independent means the quality of the data products derived from the system inputs.

National Research Council (NRC). 1995. Earth Observations from Space: History, Promise, and Reality. Washington, D.C.: National Academy Press.

National Research Council (NRC). 1998a. Overview, Global Environmental Change: Research Pathways for the Next Decade. Washington, D.C.: National Academy Press.

National Research Council (NRC). 1998b. The Atmospheric Sciences: Entering the Twenty-First Century. Washington, D.C.: National Academy Press.

National Research Council (NRC). 1999a. The Adequacy of Climate Observing Systems. Washington, D.C.: National Academy Press.

National Research Council (NRC), Space Studies Board. 1999b. "Assessment of NASA's Plans for Post-2002 Earth Observing Missions," short report to Dr. Ghassem Asrar, NASA's Associate Administrator for Earth Science, April 8.

National Research Council (NRC). 2000. Reconciling Observations of Global Temperature Change, National Academy Press, Washington, D.C.

U.S. Global Change Research Program (USGCRP). 1997. Our Changing Planet: The FY 1998 U.S. Global Change Research Program. U.S. Global Change Research Program Office, Washington, D.C.

Willson, R.C., and H.S. Hudson. 1991. The sun's luminosity over a complete solar cycle. Nature 351: 42-44.

2

Atmospheric Soundings

INTRODUCTION

The large-scale motions of the atmosphere are among the principal controls of our daily weather. These motions, referred to as atmospheric circulation, are also among the main controls of regional and global climate through the way they cause variations in cloudiness, wind, temperature, and precipitation. Atmospheric circulation is generally described in terms of three-dimensional distributions of momentum and thermodynamics, namely wind speed and direction, temperature, and moisture.

Accurate observations of global wind, temperature, and humidity are of paramount importance to numerical weather prediction (NWP) project activities that depend on the initial state defined by the three-dimensional structure of these quantities. Accurate observations of these quantities are also important for climate research because the energy budget of the climate system is governed substantially by their distribution. Although temperature and moisture profiles (soundings) are currently obtained from conventional meteorological observing networks operating over populated regions, the lack of global coverage, together with the steady demise of these networks, results in increasing reliance on satellite observations to fill critical gaps in observational data.

Setting the climate measurement requirements for temperature and moisture is a difficult task, given the integral way these parameters relate to many important climate processes. Table 2.1, extracted from the National Polar-orbiting Operational Environmental Satellite System (NPOESS) climate measurement requirements, the environmental data records (EDRs) (NOAA, 1997), summarizes sounding capabilities expected when NPOESS becomes operational. A number of questions can be raised regarding the adequacy of the stated EDR thresholds. For example, the threshold vertical resolution for humidity is unrealistic, and stated threshold accuracies reflect expected capabilities rather than actual climate needs. It remains an open but critical question whether or not the information extracted from current NWP systems, including systems planned for NPOESS, is at all sufficient to meet climate research requirements.

The committee's findings in Box 2.1 address the current status of space-based measurements and data and future needs in the integrated NPOESS program for research-quality atmospheric temperature and moisture data for the study of climate change.

TABLE 2.1 NPOESS Climate Environmental Data Record Threshold Requirements for Temperature and Moisture Soundings

System Capability	Temperature Threshold	Water Vapor Threshold (specific humidity)[a]
Horizontal Resolution		50 km
Troposphere		
Clear, nadir	50 km	
Clear, worst case	100 km	
Cloudy, nadir	50 km	
Cloudy, worst case	100 km	
Stratosphere		
Clear	200 km	
Vertical Resolution		
Clear		
1. Surface-300 millibars (mbar)	±1.0 K/1 km depth	20 mbar, surface-850 mbar
2. 300-30 mbar	±1.0 K/3 km depth	50 mbar, 850-100 mbar
3. 30-1 mbar	±1.0 K/5 km depth	
4. 1-0.01 mbar	±3.5 K/5 km depth	
Cloudy (>80%)		
5. Surface-700 mbar	±1.5 K/3 km depth	
6. 700-100 mbar	±1.5 K/3 km depth	
Measurement Accuracy	± 0.5 K	
Clear		
1. Surface-600 mbar		±25%
2. 600-400 mbar		±35%
3. 400-100 mbar		±35%
Cloudy		
4. surface-600 mbar		±25%
5. 600-400 mbar		±40%
6. 400-100 mbar		±40%
Long-term Stability		2%
Troposphere	±0.05 K/decade	
Stratosphere	±0.10 K/decade	

[a]Primarily clear-sky.
SOURCE: Adapted from NOAA (1997).

A BRIEF HISTORICAL PERSPECTIVE

A detailed history of atmospheric sounding up to 1991 is summarized in the review article of Smith (1991). The basic physics involved in the design of temperature and moisture sounders from Earth orbit was published in the late 1950s (King, 1958; Kaplan, 1959), followed by a number of papers describing different methods of retrieval (e.g., Houghton et al., 1984; Smith, 1991). The early measurements that tested these concepts were based on measurements obtained from filter radiometers with a spectral resolution ($\lambda/\Delta\lambda$) typically on the order of 100. As noted below, the next step in sounding technology is toward sounders with a much higher spectral resolution ($\lambda/\Delta\lambda \sim 1000$).

The presence of clouds in the field of view of sounders has a detrimental effect on the quality of a retrieval. The absorption properties of cloud droplets and ice particles at infrared sounding wavelengths are so strong that even thin clouds contaminate the measurement of radiances. A number of techniques have been developed to

**Box 2.1
Summary and Findings
Present Status and Future Needs of Space-Based Atmospheric Soundings**

The past 20 years have witnessed considerable progress in passive infrared remote sensing of temperature profiles using radiance data obtained from filter radiometers. Currently, the combination of the High Resolution Infrared Sounder and the Microwave Sounding Unit (MSU) provides atmospheric temperature profiles with an average root mean square (rms) error of approximately 2 K and a vertical resolution of 3 to 5 km in the troposphere. Temperature retrieval algorithms applied to data from this current suite of operational sounders are mature and well understood; however, the accuracy and resolution of temperature retrieved from current sounder data fall short of numerical weather prediction (NWP) requirements. In addition, even when identical retrieval algorithms and instruments are employed, discrepancies in the data products from one spacecraft to the next reduce the utility of the data for climate monitoring. Although new technologies such as Global Positioning System satellites are expected to improve portions of the retrieval, it is hoped that the next generation of full-column sounders will overcome the shortfalls mentioned. The situation is even worse for water vapor, where characteristics of water vapor retrievals and the accuracy of retrieved water vapor measurements are less advanced.

Issues that need to be considered in the remote sensing of temperature and moisture on the National Polar-orbiting Operational Environmental Satellite System (NPOESS) for climate research include the following:

- **Current sounding systems fall short of the requirements of numerical weather prediction.** This shortfall has prompted the development of next-generation sounders, beginning with the Atmospheric Infrared Sounder (AIRS) on the Earth Observing System (EOS) PM satellite followed by the Cross-track Infrared Sounder (CrIS) and Infrared Atmosphere Sounding Interferometer radiometer as part of NPOESS, that provide higher-spectral-resolution measurements. The traditional goal of sounding measurements is to deliver profiles of temperature and moisture for use in numerical weather prediction. These sounding data products also have direct and obvious climatological value. However, assimilation of radiance data has progressed at operational prediction centers, where explicit retrievals are no longer carried out in this traditional sense. Sounding data products are now derived as outputs from NWP models rather than as direct outputs from retrieval schemes.
- **There are additional reasons to move toward increased spectral resolution in sounders:** (1) improvements in signal-to-noise ratios can be realized by averaging channels with the same characteristic absorption; (2) clearer discrimination of thin clouds is possible using more highly resolved line absorption information; and (3) spectral measurements enhance the capability of providing other nonsounding information. Although additional advantages exist, sounding information content does not increase proportionally with increasing spectral resolution. Within this context it is legitimate to ask (1) what are the optimal placement and number of channels of a high-resolution instrument like AIRS or CrIS that actually contribute to retrieved soundings, (2) which channels are redundant, and (3) what is to be gained by combining a number of redundant channels.
- **From the climate perspective, the question of whether the calibrated radiance data obtained from sounders are more basic than the sounding products derived from these radiances is unresolved.** This is an especially relevant debate today, given the changing ways these radiance data are used in assimilation systems and the likelihood of changes in analysis systems in the future.
- **For purposes of monitoring climate change, it is critical to establish the extent to which any retrieved quantity relies on the first estimate of that quantity, which is usually derived from a climatological database.** Properties that are too dependent on such databases cannot provide proper measures of evolving climate change. Current analysis systems, especially as they apply to water vapor, are inadequate and unfortunately rely too much on existing but poorly known climatologies or time records. It is not obvious that this situation will improve substantially with the next generation of infrared sounders.

continued

> **Box 2.1 Continued**
>
> - **Radiance data, complete with calibration information, should be considered as the fundamental climate data resource in addition to retrieved sounding products.** Calibrated radiances are the basic inputs to NWP analyses and the basic inputs to retrieval algorithms. NWP analyses and algorithms have changed dramatically since the launch of the first TIROS Operational Vertical Sounder (TOVS) and may also change significantly in the future. Reprocessing the radiance data will continue to be an important activity in this environment of changing approaches to analysis.
> - **Climate measurement requirements for temperature and moisture should also be expressed in terms of radiance requirements in addition to geophysical parameters.**
> - **Maintenance and enhancement of existing, conventional observing networks, e.g., radiosonde networks, should be an integral part of the supporting and dedicated vicarious calibration efforts.** A coordinated vicarious calibration activity must accompany the satellite measurements to monitor calibration accuracy, assess differences in output from different versions of the same instrument, and determine spurious instrument-based trends. Experience has shown that proper assessment of measurement precision follows the accumulation and analysis of a sufficiently large body of data. To provide the most useful information, the conventional networks must also be supported by a program that characterizes and calibrates the instrumentation employed to ensure quality and consistency in all geographic regions.
> - **Orbit stability should be maintained to avoid aliasing errors[1] that arise from inadequate sampling of the diurnal cycle, among other errors that might affect measurement trends.**
> - **Precise intersatellite calibration, based on measurement continuity with sufficient instrument overlap, is required to determine differences between versions of the same instrument on different satellites.**
>
> ---
> [1]Errors introduced when high-frequency occurrences are interpreted as low-frequency occurrences because of inadequate sampling.

minimize the effects of clouds on soundings. These usually require some way of identifying cloudy scenes to arrive at an equivalent clear-sky radiance quantity (such as the so-called "cloud-clearing" method of Smith, 1968). It is difficult to remove the effects of clouds entirely from the data, producing larger retrieval errors under these conditions (this is also reflected in Table 2.1). Methods accounting for the effects of clouds on the data are generally based on higher-resolution visible and infrared imaging data that are required to supplement the sounding channels.

The concern over the problem of clouds motivated the inclusion of microwave sounding channels as part of the sounding instrument system. Clouds at these frequencies are almost transparent (although not transparent enough to eliminate the effects of clouds entirely). Examples of microwave sounders are the Microwave Sounding Unit (MSU) with four channels across the 57 GHz oxygen band, the Defense Meteorological Satellite Program (DMSP) Microwave Temperature Sounder (SSM/T/1) and the DMSP Microwave Water Vapor Profiler (SSM/T/2) with channels at 118 GHz and across the 183 GHz absorption line of water vapor, and the more recent Advanced Microwave Sounding Units A and B (AMSU-A and AMSU-B).

Although this section addresses soundings from the perspective of polar orbiters, the same observations apply to instruments in geosynchronous Earth orbit (GEO). For example, the GEO instruments (including Polar-orbiting Operational Environmental Satellites, POES) have in general lacked the precise characterization capability and stability necessary for later reconstruction of long-term climate data sets. Additionally, the sheer volume of GEO data has created difficulties for scientists who wish to access the data for postprocessing. Finally, because specific sectors of the globe are monitored for largely national missions, national interests come into play when decisions regarding sharing of data arise. Nevertheless, data from GEO sources, with their unique simultaneous "full-disk" observations of the planet, should be considered a part of the overall strategy for monitoring climate variation. Such data can be used to cross-calibrate POES data and fill in the gaps in the typical POES swath.

OBSERVING STRATEGIES

Temperature and moisture profiling relies on spectral measurements of radiation emitted by absorbing gases in the atmosphere. If the distribution and absorption of a gas are known, as they are for the uniformly mixed gases of CO_2 and O_2, then the detected emission is proportional to the temperature of that volume of gas. For other gases such as H_2O that are not uniformly mixed, the emission depends on both temperature and the concentration of the gas itself. Therefore, profiling water vapor requires having simultaneous data on temperature, a requirement that complicates the process.

The strength of the absorption by the gas more or less determines the levels at which the emission occurs and provides a way of profiling information as a function of height. Emission in strongly absorbed regions occurs higher in the atmosphere than does emission at weakly absorbed wavelengths. This property is characterized by the contribution, or weighting, function. The shape of the weighting function, important because it essentially defines the vertical resolution of the sounding measurements, depends on a number of factors. There are two primary factors: (1) vertical distribution of the absorbing gas—or, stated differently, the scale height of the atmosphere, which establishes the upper limit to the resolution, which is approximately 1 km, and (2) spectral resolution of the measurement, which also affects the vertical resolution. An instrument that effectively averages over many wavelengths smears out the individual weighting functions of each wavelength, producing a more broadened function and reducing the resolution. Increasing spectral resolution will increase vertical resolution, but only to the point of the upper limit. An unavoidable consequence of the broad nature of the weighting functions is that a function at one (spectral) channel significantly overlaps the functions associated with adjacent (spectral) channels. This overlapping property is the source of the two most significant problems in sounding retrievals: (1) the measurements are not entirely independent, leading to inverse solutions that are not unique, and (2) the inversions are unstable, with small errors (measurement and other) producing large changes to the solutions.

Sounding Product Issues

For the reasons mentioned above, limiting solutions to some a priori or initial guess is a necessary step in obtaining meaningful profiles of temperature and moisture. This constraint arises from some profile information obtained from a climatological database of one sort or another. Errors introduced by these limitations are not a major concern if it is known that initial estimates do not propagate into the final retrieval. For characterizing climate and especially for monitoring change, it is thus critical to establish the extent to which any retrieved quantity relies on an initial estimate, which is usually derived from an unreliable climatological database. If the analysis of any properties is too dependent on such information, it will not provide proper measures of evolving climate change. Unfortunately, water vapor information obtained from current sounders relies heavily on a priori data (Engelen and Stephens, 1998). It is not obvious that the water vapor information derived from the new sounders being developed for use on NPOESS and other future platforms will alter this situation.

Another important factor in the retrieval of moisture soundings concerns the accuracy of the forward model used to simulate the measured spectral radiances. The major source of error in water vapor retrievals does not stem from radiance calibration errors but rather from errors of the forward model. These errors remain large, and efforts are now under way to establish some understanding of their nature.

Radiance Data Product Issues

Operational sounders have provided continuous, quasi-global data since the late 1970s with the launch of the first version of the TIROS Operational Vertical Sounder (TOVS). Radiance data from multiple versions of the TOVS flown on a series of National Oceanic and Atmospheric Administration (NOAA) polar-orbiting spacecraft have been combined to produce radiance climatological databases, or climatologies. Two examples of these radiance climatologies are found in the work of Spencer and Christy (1992) and Bates et al. (1996). These studies pointed to a number of common issues that arise when combining data from different satellites to produce a climatology:

1. To establish the precision of the long time series of radiance data required for identifying trends in data records, it will be necessary to identify and remove biases that arise from (a) incomplete or changing sampling practices throughout the data record, (b) changes in instrument orientation (drifts), and (c) measurement differences between different versions of the same instrument. Sufficient overlap in the measurement obtained with different versions of a given instrument can be used to determine the magnitude of measurement differences and thereby to remove biases in the data. An approach often used to assess the likely magnitude of resulting errors that arise from effects of diurnal aliasing is to sample data from geostationary satellites (e.g., Salby and Callaghan, 1997).

In addition to the problem of temporal sampling, the NOAA spacecraft that carry the TOVS are placed in a nominal Sun-synchronous orbit. Unfortunately, these orbits drift at a disturbing rate. Orbital drifts and decays create spurious trends in the data (e.g., Wentz and Schabel, 1998). In addition to drifts, there have been periods of time since 1981 when data from only one satellite were available, which is a further source of bias associated with inadequate sampling of the diurnal cycle. Both Bates et al. (1996) and Christy et al. (1995) introduced procedures to account for biases that arise from intersatellite sensor differences.

2. To determine the accuracy of radiance data over a protracted period of time, a comparison of multiple measurements is essential. Reliance on prelaunch calibration is questionable, given the usual changes in instrument responsivities in orbit. The method of vicarious calibration, such as that used routinely to calibrate geostationary water vapor channels (e.g., Van de Berg et al., 1995), is required in addition to on-orbit calibration. Difficulties in vicarious calibration of water vapor channels arise through the lack of accuracy of radiosonde data on upper tropospheric water vapor. This data inaccuracy limits the ability to match measured and simulated radiances. Christy et al. (1995) provided an example of a vicarious calibration approach that they used to assess the precision of the MSU radiances. They compared MSU temperature variations simulated from radiosonde data with actual measured temperatures.

3. Radiance data ultimately require some form of interpretation relative to more conventional climate parameters. Much of the controversy associated with analyses of radiance trends has to do with the interpretation of the data.

EVOLUTION STRATEGY

At the time of the Global Weather Experiment in 1979, it was hoped that the use of satellite-derived temperature and moisture profiles would extend the range of useful synoptic-scale forecasts beyond a few days to a week or more. Although NWP systems have improved over the last decade, it has become increasingly difficult to demonstrate that temperature profiles retrieved from satellite sounding data have a consistent positive impact on Northern Hemisphere forecasts (e.g., Eyre et al., 1992; Smith, 1991). This difficulty has led to the belated recognition that the information content of temperature data obtained from current sounders is low, relative to temperature "knowledge" already contained in NWP systems. It is argued that low information content and thus minimal impact on forecasts result from the poor vertical resolution of the data. The next-generation satellite sounders, beginning with the Atmospheric Infrared Sounder (AIRS) on NASA's Earth Observing System Afternoon Satellite (EOS PM), are expected to provide information at higher vertical resolution and thus presumably have a more positive influence on forecasts.

Evolution to Higher Spectral Resolution

There are a number of reasons to move toward increased spectral resolution in sounders: (1) improvements in the signal-to-noise ratio can be realized by averaging channels with the same characteristic absorption; (2) clearer discrimination of thin clouds is possible using more highly resolved line absorption information; and (3) spectral measurements enhance the capability of providing other nonsounding information, for example, about clouds and particle sizes. Given these advantages, however, sounding information content does not increase proportionally with increasing spectral resolution. Within the context of AIRS measurements of temperature, analyses of the significance of the information (e.g., Twomey, 1996; Rodgers, 1996) revealed that the AIRS spectra contained about 14 pieces of independent information on vertical temperature profiles, which translates roughly to the 1 km vertical resolution limit noted previously.

Therefore, it is legitimate to ask which channels of a high-resolution instrument such as AIRS or the Cross-track Infrared Sounder (CrIS) optimally contribute to retrieved soundings, which channels are redundant, and what is to be gained by combining a number of redundant channels (Rodgers, 1996).

Evolution to Assimilation of Radiance Data

As an understanding of the true information content of sounding data emerged in the 1990s, alternate approaches were developed to account for it and for the error characteristics of the data. These methods are based on the assimilation of the radiance data into the NWP system, which could weight the information more appropriately based on the proper error characteristics of the data (Eyre and Lorenc, 1989). Assimilation of radiance data has produced a clear, positive impact on NWP. For example, McNally and Vesperini (1996) showed how assimilating TOVS radiance data in the European Center for Medium-Range Weather Forecasts significantly improves the analysis of many aspects of the hydrological cycle, leading to better forecasts.

CHALLENGES AHEAD

In the near future, new, more highly resolved spectral measurements promise to improve researchers' ability to profile temperature in the atmosphere and to improve the present inadequate techniques for profiling water vapor. How the sounding data contribute to observing and understanding Earth's climate will depend in part on whether current instrumentation can meet the EDR requirements (see Table 2.1) for the vertical resolution proposed for water vapor.

Traditionally, sounding data have delivered profiles of temperature and moisture for use in numerical weather prediction. These sounding data products have direct and obvious climatological value. However, radiance assimilation at operational prediction centers no longer carries out explicit retrievals in the traditional sense. These sounding products are now derived as outputs from NWP models. Whether the optimal use of new-generation, higher-spectral-resolution sounding data will continue the present trend of using radiance data directly is still an open issue. Nevertheless, the value of carefully calibrated radiance data not only for assimilation purposes, but also as a resource for climate data, clearly emerges from all current studies.

REFERENCES

Bates, J.J., X. Wu, and D.L. Jackson. 1996. Interannual variability of upper-tropospheric water vapor band brightness temperature. J. Climate 9: 427-438.

Christy, J.R., R.W. Spencer, and R.T. McNider. 1995. Reducing noise in the MSU daily lower-tropospheric global temperature data set. J. Climate 8: 888-896.

Engelen, R., and G.L. Stephens. 1998. Characterization of water vapour from TOVS/HIRS and SSMT-2 measurements. Q.J.R. Meteorol. Soc. 125: 331-351.

Eyre, J.R., and A. Lorenc. 1989. Direct use of satellite sounding radiances in numerical weather prediction. Meteorol. Mag. 118: 3-16.

Eyre, J.R., E. Andersson, and A.P. McNally. 1992. Direct use of satellite sounding radiances in numerical weather prediction. NATO ASI Series 9: 365-380.

Houghton, J.T, F.W. Taylor, and C.D. Rodgers. 1984. Remote Sounding of Atmospheres. New York: Cambridge University Press.

Kaplan, L.D. 1959. Inferences of atmospheric structures from satellite radiance measurements. J. Opt. Soc. Am. 49: 1004.

King, J.I.F. 1958. The radiative heat transfer of planet Earth. Scientific Uses of Earth Satellites, 2nd revised Ed. Ann Arbor: University of Michigan Press.

McNally, A.P., and M. Vesperini. 1996. Variational analysis of humidity information from TOVS. Q.J.R. Meteorol. Soc. 122: 1521-1544.

National Oceanic and Atmospheric Administration (NOAA). 1997. Climate Measurement Requirements for the National Polar-orbiting Operational Environmental Satellite System (NPOESS): Workshop Report, Herbert Jacobowitz (ed.), Office of Research Applications, NESDIS-NOAA, February. 77 pp.

Rodgers, C.D. 1976. Retrieval of atmospheric temperature and composition from remote measurements of thermal radiation. Rev. Geophys. Space Phys. 14: 609-624.

Rodgers, C.D. 1996. Information content and optimization of high spectral resolution measurements. SPIE 2830: 136-147.

Salby, M., and P. Callaghan. 1997. Sampling error in climate properties derived from satellite measurements: consequences of undersampled diurnal variability. J. Climate 10: 18-36.

Smith, W.L. 1968. An improved method for calculating tropospheric temperature and moisture from satellite radiometer measurements. Mon. Weather Rev. 96: 387-396.

Smith, W.L. 1991. Atmospheric sounding from satellites. Q.J.R. Meteorol. Soc. 117.

Spencer, R., and J. Christy. 1992. Precision and radiosonde validation of satellite grid point temperature anomalies, Part I: MSU channel 2. J. Climate 5: 847-857.

Twomey, S. 1996. Introduction to the Mathematics of Inversion in Remote Sensing and Indirect Measurements. New York: Dover.

Van de Berg, L.C.L., J. Schmetz, and J. Whitlock. 1995. On the calibration of the Meteosat water vapor channel. J. Geophys. Res. 100: 21069-21076.

Wentz, F., and M. Schabel. 1998. Effects of orbital decay on satellite derived lower tropospheric temperature trends. Nature 394(6694): 661-664.

3

Sea Surface Temperature

INTRODUCTION

Sea surface temperature (SST) has been used for centuries as a way to trace the origin of surface waters and gain greater knowledge of a location—for example, Benjamin Franklin's maps of the Gulf Stream were used to speed mail between England and North America. High-quality SST fields have been archived since the late 19th century. Surrogate observations from coral isotopic ratios, for example, allow scientists to extend records back several hundred years into the past in some parts of the world's ocean. This makes SST one of the more robust indicators for understanding Earth's climate. The ability to monitor global and regional surface temperature has improved so that it is now possible to use SST observations as indicators of regional- to basin-scale change, as well as for forecasting stress on the natural flora and faunal assemblages.

The National Polar-orbiting Operational Environmental Satellite System (NPOESS) environmental data record (EDR) requirements and goals for SST are aggressive, and their achievement would facilitate the utility of NPOESS observations for climatic research purposes, but attention to the entire observing process is necessary to make such fields generally useful.

This chapter reviews the underlying scientific issues, current and future directions, observing strategies, and concomitant needs for processes such as calibration, validation, and data management to maximize the utility of satellite observations of SST. The committee's findings in Box 3.1 address the present status of space-based measurements and data, as well as future needs in the integrated NPOESS program for research-quality SST data in the study of climate change.

BASIC SCIENCE ISSUES

Observation of SST in the modern era started with capturing buckets of seawater from over the sides of ships, immersing a mercury thermometer into each bucket, and recording the water temperature. Such measurements were made widely from merchant ships in the 18th and 19th centuries and are still made today. These point measurements were widely separated in time and space, and observers tended to collect seasonal assemblages of measurements and produce large-scale analyses (maps) or thermal analyses based on the assumption that the ocean was a very slowly changing medium. By the mid-20th century, however, it was clear that many parts of the ocean change quickly enough to invalidate such an assumption. In response, observers have tried to increase the density

> **Box 3.1**
> **Summary and Findings**
>
> There are two substantive issues associated with the NPOESS environmental data record (EDR) requirements for sea surface temperature (SST) observation. First, the SST requirement has not been modified to be compatible with the results of the Climate Requirements Workshop Report (CRWP) (Jacobowitz et al., 1996). Second, calibration and validation are not a part of the NPOESS specification.
>
> The CRWP developed a modified set of ocean observation requirements for SST. It noted that the accuracy objective can be relaxed from 0.1 K[1] at pixel resolution to 25 km spatial scale and a one-week temporal scale. It also commented on resolution of diel[2] effects and instrumental stability. The CRWP suggested that resolution of the diel cycle would require a constellation of four polar platforms, rather than the three specified by NPOESS, i.e., 3-hour temporal sampling. The CRWP recommended that satellite-based instruments have a demonstrable stability of 0.1 K.
>
> Monitoring stability in the SST record during NPOESS missions and in the handoff periods between NPOESS and Earth Observing System (EOS) platforms requires extensive preflight characterization and post-launch validation, as outlined in the text of this chapter. The NPOESS EDRs neither specify such an activity nor suggest how sensor providers will demonstrate on-orbit stability that meets the requirements. On a more general level, there is no strategy either to integrate the lessons learned from EOS into NPOESS, or to provide inter- or intrasystem validation.
>
> ---
>
> [1] In space studies of weather and climate, it is customary to denote temperature in the Kelvin, or absolute, scale. For Celsius temperature readings, 0 °C = 273.16 K.
> [2] Diel is defined as a variation over a 24-hour period.

of ship platforms, develop autonomous floats, and use aircraft and spaceborne radiometric sensors to obtain more synoptic views of the surface thermal field.

Interestingly enough, the increasing density of observations has shown that each approach provides a slightly different estimate of the SST because of the peculiarities of each sampling system. This understanding has motivated development of techniques for assimilation that attempt to compensate for such peculiarities and provide surface temperature fields with known characteristics.

Generally, SST analyses have been of two types: pattern discrimination and quantitative field estimates. It should be noted that satellite infrared (IR) observations of surface temperature were initiated to support meteorological applications, not to further oceanographic or climatic purposes; later such requirements started to drive accuracy, spectral placement of radiometer windows, and so on. Early satellite-based analyses could discern the edge of the Gulf Stream or the California Current because of the strong surface temperature gradient. However, the estimate of surface temperature might have been accurate only to 1.5 K or so. Eddies, boundary currents, and other mesoscale phenomena were readily identified, even though the accuracy of the estimated temperatures in and around the features was less than that obtained with in situ techniques. Moreover, the level of accuracy was not useful for following large-scale, low-frequency temperature change in the surface ocean, such as might be caused by the El Niño/Southern Oscillation (ENSO), the North Atlantic Oscillation, or other lower-frequency phenomena. During the last two decades, however, the accuracy of SST mapping from satellites has improved so that observations are routinely produced at rms (root mean square) accuracies of 0.6 K or better, thus permitting the observation and study of large-scale, low-frequency fluctuations in the ocean that might be associated with climatic variation or ecosystem change.

Current Applications

History plays a role in current applications of satellite SST analyses. Development of algorithms for producing reliable SST data sets from spaceborne infrared radiometers has been pursued by a number of investigators, agencies, and governments since the late 1960s (see reviews by Brown and Cheney, 1983; Abbott and Chelton, 1991). For example, the National Oceanic and Atmospheric Administration (NOAA) (McClain, 1981; McClain et al., 1983; Strong and McClain, 1984; McClain et al., 1985), National Aeronautics and Space Administration (NASA) (Shenk and Salomonson, 1972; Chahine, 1980; Susskind et al., 1984), and Rutherford Appleton Laboratory in the United Kingdom (RAL/UK) (Llewellyn-Jones et al., 1984) have addressed infrared radiometry using a variety of radiation transfer codes, modeled and observed vertical distributions of temperature and moisture, and actual observations. Minnett (1986, 1990) and Barton et al. (1989) summarized the present state of the art for high-quality retrievals from NOAA's Advanced Very High Resolution Radiometer (AVHRR)-class instruments. The current state of the art is limited by radiometer spectral interval placement, radiometer noise, quality of prelaunch instrument characterization, in-flight calibration quality, viewing geometry, atmospheric correction, and characterization of the quality of the SST retrieval, such as contamination by cloud and aerosols.

As noted above, SST analyses are generally based on temperature estimates and identification of patterns. Thermal field estimates are used to initialize climate models as well as for diagnostic applications, for example, determining if a boundary current is in the correct location, or if the estimated seasonal temperature range in a region is correct. Similarly, molecular and thermal fluxes have a significant dependence on SST. The ocean-atmosphere CO_2 exchange is strongly influenced by SST (Van Scoy et al., 1995), while explosive cyclogenesis off the U.S. East Coast has been associated with the warm waters of the Gulf Stream system (Raman and Niyogi, 1998).

ENSO Studies

Tracking equatorial Pacific SST anomalies has become a pastime as ENSO impacts have risen in the public consciousness, and maps of equatorial Pacific SST anomalies have become commonplace in the media.[1] From a scientific perspective, SST fields are used to initialize and validate coupled tropical models.

Surface Current Pattern and Magnitude

Pattern tracing of SST gradients for boundary currents and associated mesoscale structure is used daily by the U.S. Navy, NOAA, and private enterprise for maritime defense and commercial applications. These applications include optimal ship routing, fisheries catch management, antisubmarine warfare, and petroleum exploration and exploitation. Scientific applications fuse other sensor systems, such as altimetry, to obtain estimates of surface layer velocity and transport (Kelly and Gille, 1990; Gõni et al., 1997).

Nutrient Estimates

The correlated nature of upper-mixed-layer thermal and nutrient fields facilitates the use of SST fields as proxies for areal nutrient concentrations (Kamykowski, 1987). This augments in situ observations needed for model initialization and validation and provides researchers with an approach to determining large-scale nutrient field variations unobtainable by any other means.

Coral reef assemblages are sensitive to ambient water temperature, responding to extended periods of abnormally warm water by expelling symbiotic algae. Satellite SST maps have been used to study eastern Pacific episodes of such "bleaching" (Podestá and Glynn, 1997), while NOAA has forecast large-scale impacts on the Great Barrier and Florida Keys reef systems (Montgomery and Strong, 1994). Satellite SST fields are cost-effective sentinels of these events.

---

[1]See, for example, the information available online at <http://www.pmel.noaa.gov/toga-tao/gif/daily/sst_wind_anom_5day.gif>.

Mesoscale Studies

During the 1970s, oceanographers started using satellite SST imagery to assist experimental logistics and data analysis in the Somali, California, and Gulf Stream systems (Brown and Evans, 1980, Brown et al., 1985; Abbott and Zion, 1987). The daily or semidaily synoptic coverage makes satellite SST ideal for tracking changes in the oceanic mesoscale, weather target eddies, boundary currents, and squirts and jets. Strategies to determine boundary current loci (Cornillon and Watts, 1987; Olson, 1991; Gõni et al., 1997) have used satellite SST and altimetric systems to good effect. Eddy tracking by SST has been used in all western boundary current systems. Smaller-scale studies of estuarine variability (Framiñan and Brown, 1996) are also facilitated by satellite observations.

FUTURE DIRECTIONS

Polar-orbiting satellite observing systems sense ocean parameters daily, at best. Cloudiness, sensor swath gaps, and orbital mechanics limit observation frequency. Global observations have also been constrained by sensor, telemetry, and ground-processing technologies. The next decade will see multiple satellite platforms carrying similar observing technologies in Sun-synchronous orbits at differing local times of day (Sun time). Sampling from this combined system should improve retrievals from cloudiness-limited scenes and permit the resolution of diel variations in clear areas. However, assimilation of observations at differing local Sun times necessitates improved understanding of diel cycles in the parameters of interest. For example, current observations from NOAA AVHRR platforms show significant changes in satellite SST fields as the crossing time drifts later in the day, as one might expect. Geosynchronous platforms now provide subdiel temporally resolved SST fields. Innovative compositing and assimilation approaches can remove much of the data contamination caused by clouds, but here, too, the enhanced sensitivity of the instruments now resolves diel effects. Similarly, use of passive microwave techniques to discern SST variation in predominantly cloudy areas will improve overall understanding of tropical and polar air-sea exchange processes.

Coastal Processes

High-frequency variability in time and space is typical of nearer-shore processes. Tidal effects, subinertial oscillations, coastal trapped waves, and coupled land-sea effects all contribute to a rich high-frequency spectrum. Significant improvements in temporal and spatial sampling are needed to resolve many coastal processes adequately. Temporally, this indicates a need for geosynchronous Earth orbit (GEO) sensors, while spatially, low Earth orbit (LEO) sensors have the advantage. Observation of coastal processes will probably demand linking of GEO and LEO observations into coherent depictions of phenomena. Such higher-frequency fields would complement systems currently used by linking the higher-frequency variance to the larger scales.

Skin Temperature

Much progress has been made in the past several years on in situ sensing of skin, or surface radiation, temperature. New technologies applied to this problem (Smith et al., 1996) have provided high accuracy, improved coverage, and diel-resolved observing sets. One can expect that in the future most infrared instrumentation will be validated by comparison with surface radiation temperature, rather than bulk temperature. Bulk-skin temperature differences tell much about the surface heat flux, a topic of major interest to regional and global climate studies.

Upper-Ocean Mixed-Layer Temperature

A combination of satellite-sensed SST with surface wind observations and in situ profile observations meets minimal requirements for initializing and validating upper-ocean mixed-layer models. Such model systems provide a dynamic framework for the assimilation of satellite and in situ observations and the forecasting of mixed-layer evolution. This will probably be one of the most important applications for satellite SST observations in the early years of the 21st century. It is also an ideal framework for implementing regional and basin-scale observing systems.

OBSERVING STRATEGY

Requirements for SST observations to meet climate, national defense, and science needs have been articulated by a number of groups (WOCE, 1985; IPO NPOESS, 1996; NOAA, 1997). NOAA identified a set of SST observation objectives (NPOESS, 1995) as part of its requirements development process for the converged NPOESS. These objectives push the state of the art in accuracy and temporal and spatial resolution (0.1 K, 0.5 km, 3-hour revisit). They are excellent goals, but achieving them would necessitate a fully operational four-satellite constellation with a cloudless Earth, state-of-the-art radiometers, and a robust data assimilation and validation system. Clearly, such an observing system would fulfill many of the needs articulated in the previous section. However, maintaining such a system would require a robust prelaunch characterization program, postlaunch validation, appropriate data assimilation procedures, and continuing quality assurance of system performance.

The major challenge in the short term will be to develop a systematic approach to the use of satellite IR and microwave radiances and derived temperatures from disparate platforms. International collaboration on such systems is recent,[2] and the current and nearer-term sensor systems have not been well coordinated. For example, one may find several platforms in orbits with similar overpass times, or systems with a similar SST product but different observing channels, or systems with quite different atmospheric correction approaches. Any one of these issues poses a challenge; we will probably see all of them during the next decade.

NASA's and NOAA's Plans

During the next 5 years SST observations will be produced by NOAA and the Navy from NOAA's AVHRR system, by NASA from the EOS/MODIS, by the National Space Development Agency (NASDA–Japan) from ADEOS-2/GLI, and by the European Space Agency (ESA) from ENVISAT/Advanced Along Track Scanning Radiometer (AATSR). This is a minimum estimate; other investigators and nations may also develop sensors and analysis systems. Table 3.1 lists some characteristics of known systems that should be in orbit in the next 5 years. Table 3.1 covers passive IR systems; there are also several passive microwave systems being put in place whose composite performance characteristics approach those of the IR systems.

TABLE 3.1 Comparison of Low Earth Orbit Infrared SST Observing Systems (1999-2004)

System/Attribute	NOAA AVHRR	EOS MODIS	ADEOS GLI	ENVISAT AATSR
Local Sun time	0730/1430	1030	1030	1000
Spatial resolution (km)	1	1	1	1
Approach	NLSST	NLSST	MCSST	Dual Path/MCSST
NEΔT (K)	0.12	0.05	0.10	0.025
Validation approach	Bulk	Skin	Bulk	Skin
Expected SST accuracy (K)	0.55	0.40	0.50	0.30

NOTE: Acronyms are defined in Appendix B.

---

[2]Committee on Earth Observation Satellites (CEOS). 1997. Towards an Integrated Global Observing Strategy, Strategic Implementation Team Scoping Paper. Available online at <http://www.eos.co.uk/ceos-calval/igos/sitscope.htm>. Committee on Earth Observation Satellites (CEOS). 1998. Final Report of the CEOS Analysis Group 1996/1997). Available online at <http://www.smithsys.co.uk/IGOS/Agfinalreport.html>.

Variations in observing times and validation approaches suggest that the first aspect of an observing strategy must be to understand the diel SST (and skin-bulk temperature) cycle and use such understanding to analyze satellite SST estimates. Second, prior to launch all these systems must be cross-referenced to common standards and characterized by similar approaches. The differing validation approaches suggest that a multinational group should coordinate the joint validation of these missions. Since development of atmospheric correction approaches uses radiative transfer calculations, a common suite of tools should be developed and provided to interested platform operators.

Mean cloudiness at midlatitudes varies with season and hemisphere; overall cloud-contaminated pixels can exceed 60 percent over much of the year (R. Evans, Rosenstiel School, University of Miami, personal communication). Strategies to fill in data gaps caused by clouds or other phenomena span the range from linear interpolation to Laplacian relaxation techniques (i.e., techniques involving a differential operator that identifies a particular type of mathematical interpolation), to objective analysis and dynamical interpolation, to blending of IR and microwave-based observations. There are differing approaches to objective analysis and dynamical interpolation using estimated satellite SSTs, brightness temperatures, or radiances. Similarly, we have no validated blending approach for IR and microwave observations. There is no standard approach to solving the "gappiness" problem, which may be caused by temporal sampling characteristics of a system, cloudiness, or a variety of other factors.

SST fields are usually tailored for their intended use. For example, weather forecasting requires the best estimate of the current state of the SST fields with a narrow time window, while climate applications might use weekly to monthly fields. An ocean logistical or experimental application might rely on a collection of individual satellite passes. There is no single set of requirements for temporal and spatial resolution, level of accuracy, and acceptable degree of error.

Nationally, four operational SST products are obtained using satellite IR observations: an analysis from the Fleet Meteorology and Oceanography Center (FMOC) (Clancy et al., 1992; Cummings et al., 1997); a Naval Oceanographic Office (NAVO) analysis (May et al., 1998); a NOAA Climate Analysis Center (CAC) objective analysis (Reynolds and Smith, 1994); and a NOAA National Environmental Satellite, Data, and Information Service (NESDIS) MCSST analysis (McClain et al., 1983). These products are contrasted in Table 3.2.

Most importantly, all these analyses are based on different approaches, use somewhat different data, and are validated against bulk measurements. Each product is produced for particular applications, for example, fleet operations or climate studies, and there are clear differences among them (R. Evans, University of Miami, personal communication). Currently none of these products uses passive microwave radiances for SST estimation, but preliminary results from the Tropical Rainfall Measuring Mission suggest that a composite infrared-microwave approach may offer improvements for tropical SST estimation (F. Wentz, Remote Sensing Systems, personal communication).

Current SST products depend strongly on the characteristics of the sensor and the atmospheric correction approach employed. Most atmospheric correction approaches are combinations of brightness temperatures, either linear or nonlinear, with weighting coefficients. Coefficients are determined by regression analysis of the foregoing features versus surface observations or from forward radiative transfer calculations. None of the standard

TABLE 3.2 Comparison of National Operational SST Analysis Products

Product/Attribute	NAVY FMOC	NAVY NAVO	NOAA CAC	NOAA NESDIS
Approach	Model (OTIS)	Range checks	Optimal interpolation	Range checked
Spatial resolution (km)	111 (20 regional)	9	111	50 (14 near United States)
Temporal resolution	Daily	Daily	Weekly	Daily
Contents	Satellite/ship/drifter	Satellite	Satellite/ship/drifter	Satellite
Validation approach	Bulk	Bulk	Bulk	Bulk

NOTE: Acronyms are defined in Appendix B.

approaches adequately addresses aerosol effects, although the slant path approach used by the Along Track Scanning Radiometer (ATSR) is superior to other techniques. Aerosols remain a problem due to a lack of appropriate characterization and sufficient channels to determine their properties in the satellite data. Upcoming instruments such as MODIS include channels to sense aerosol radiance in the near-IR, so it is possible that substantial progress will be made on this problem over the next few years (see Chapter 7 of this report for a detailed discussion of aerosols).

IPO/NPOESS Requirements

The EDR requirements for SST observations developed as part of the Integrated Program Office (IPO)/ NPOESS *Integrated Operational Requirements Document*, first version (IORD-1) (IPO NPOESS, 1996) (Table 3.3) are contrasted with NOAA/AVHRR and EOS/MODIS in Table 3.4.

From a review of Table 3.4, it is apparent that many of the NPOESS SST requirements are similar to the capabilities of AVHRR and the objectives of MODIS. Some are clear improvements, such as the measurement accuracy and precision objectives. IORD-1 requirements suggest a measurement capability marginally more accurate than that of AVHRR, but with improved pixel navigation, measurement precision, and long-term stability. On the other hand, attainment of the NPOESS objectives would imply an instrument with measurement accuracy and precision exceeding those of the Earth Resources Satellite (ERS)-1/2 ATSR instrument (ATSR accuracy, approximately 0.3 K; precision, 0.1 K, Mutlow et al., 1994; Harris et al., 1995; Mason et al., 1996).

Some reviewers of the IORD-1 requirements believe that such levels of accuracy may not be attainable from a satellite-based SST observing system (Jacobowitz et al., 1996).

TABLE 3.3 NPOESS Environmental Data Record Requirements for Sea Surface Temperature (IORD-1)

Systems Capability	Threshold	Objective
Horizontal Resolution		
1. Global, nadir	3 km	1 km
2. Global, worst case	4 km	TBD[a]
3. Regional, nadir	1 km	0.25 km
4. Regional, worst case	1.3 km	TBD
Horizontal Reporting Interval		
Horizontal Compliance		
Measurement Range	271 to 313 K	271 to 313 K
Measurement Uncertainty (rms)	0.5 K	0.1 K
Measurement Accuracy	0.2 K	0.1 K
Measurement Precision	TBD	0.1 K
Mapping Uncertainty		
1. Global, nadir	1 km	0.5 km
2. Global, worst case	3 km	TBD
3. Regional, nadir	1 km	0.1 km
4. Regional, worst case	3 km	TBD
Maximum Local Average	6 h	3 h
Maximum Local Refresh	TBD	TBD

SOURCE: Extracted from IPO NPOESS (1996). The updated IORD and other documentation related to the NPOESS program are available online at <http://npoesslib.ipo.noaa.gov/ElectLib.htm>.
[a]TBD, to be determined.

TABLE 3.4. Comparison of NPOESS Sea Surface Temperature Requirements (IORD-1) with AVHRR Capabilities and MODIS Systems Objectives

Systems Capabilities	NPOESS Thresholds	NPOESS Objectives	AVHRR Actual	MODIS Objectives
Horizontal Resolution				
1. Global, nadir	3 km	1 km	4 km	4 km
2. Global, worst case	4 km	TBD[a]	4 km	4 km
3. Regional, nadir	1 km	0.25 km	1 km	1 km
4. Regional, worst case	1.3 km	TBD	1 km	1 km
Mapping Uncertainty				
1. Global, nadir	1 km	0.5 km	2 km	0.1 km
2. Global, worst case	3 km	TBD	6 km	0.1 km
3. Regional, nadir	1 km	0.1 km	2 km	0.1 km
Measurement Range	271 to 313 K	271 to 313 K	271 to 313 K	271 to 313 K
Measurement Uncertainty (rms)	0.5 K	0.1 K	0.7 K	0.4 K
Measurement Accuracy	0.2 K	0.1 K	0.2 K	0.1 K
Maximum Local Average	6 h	3 h	6 h	12 h

[a]TBD, to be determined.

CALIBRATION AND VALIDATION

Validation of SST measurements is required over the lifetime of the satellite mission. The validating instruments must be deployed in situations that encompass the entire range of surface temperatures and atmospheric variability. Since no single approach provides a perfect validation measurement, a selection of techniques and instruments is required for an adequate validation data set. The approach includes validation of (1) top-of-the-atmosphere radiances, (2) surface radiances, and (3) surface temperatures.

There are three possible methods of validating top-of-the atmosphere radiances:

1. Comparison with other satellite measurements,
2. Comparison with aircraft radiometers flying at lower elevations than the satellite, and
3. Use of radiative transfer modeling to simulate satellite measurements.

Comparison with Other Satellite Measurements

Surface radiance measurements are usually validated using calibrated spectroradiometers, such as the Marine-Atmospheric Emitted Radiance Interferometer (M-AERI) (Smith et al., 1996), or broadband infrared thermometers. These instruments can be mounted on low-flying aircraft (Saunders and Minnett, 1990; Rudman et al., 1994; Smith et al., 1994), ships (Schluessel et al., 1987; Smith et al., 1996), or fixed platforms.

Comparisons of top-of-the-atmosphere measurements obtained with satellite-borne infrared radiometers on different spacecraft have the advantage over the airborne or surface-based instruments described above of comparing the results of measurements by similar instruments. The problems with this approach are (1) the possible changes in the top-of-the-atmosphere radiation field between the two satellite overpasses (resulting from changes in the surface temperature or in the intervening atmosphere), (2) differences in the viewing geometry of the two satellites, (3) differences in the spectral responses of the different satellite instrument channels, and (4) possible less-than-required accuracy level or noise characteristics of the validating instrument.

Another possible problem is the potential for undetected in-flight degradation of the validating radiometer. If systematic discrepancies are found, it may not be apparent which satellite sensor is at fault.

Comparison with Aircraft Radiometers

A significant advantage of using aircraft radiometers is that data can be taken simultaneously with satellite measurements. However, because of the difference between spacecraft and aircraft speeds, few truly coincident measurements can be made, although within, for example, a 30-minute window of the satellite overpass a large number of validation measurements could be obtained, depending on the time interval selected for data acquisition (Minnett, 1990). Also, aircraft radiometers can in principle be arranged to match the satellite viewing geometry and can be scheduled (again, in principle) to avoid conditions that would make data interpretation difficult (e.g., broken cloud fields).

Disadvantages of this technique include the effects of the atmosphere above the aircraft, which can be accounted for by modeling, using an assumed (or measured) temperature and humidity profile, and the accuracy of the aircraft instruments. Candidate aircraft instruments for top-of-the-atmosphere radiance validation of the channels used in SST determination include the MODIS Airborne Simulator (MAS) (King and Herring, 1992) and the High-resolution Interferometer Sounder (HIS) (Bradshaw and Fuelberg, 1993). These instruments are flown typically on the NASA ER-2 research aircraft at a height of approximately 20 km, and under these conditions the spatial resolution is 50 m (MAS) and 2 km (HIS).

The noise levels of these instruments are not as low as those for the MODIS infrared channels. For the MAS, the NEΔT (or noise-equivalent delta radiance) is approximately 0.3 K for a target at about 290 K for the 3.7 to 4.0 µm channels and 0.1 to 0.2 K for the 11 to 12 µm channels. However, these levels could be greatly improved (by a factor of 20 if the noise were truly random) by averaging the data down to a typical spatial resolution of about 1 km^2. The noise levels in the HIS spectra in the 800 to 1050 cm interval are typically 0.2 to 0.45 mW m^{-2}st^{-1}cm, and these result in an uncertainty of about 0.15 K in the skin SST retrieved from the HIS spectra (Nalli, 1995).

Use of Radiative Transfer Models

The use of numerical models of radiative transfer through the atmosphere to simulate satellite measurements requires high-quality measurements of the relevant atmospheric properties (temperature and humidity profiles, aerosol characteristics) and emitted radiance at the surface, taken at the time of the satellite overpass. The advantage of this approach is that a large database of measurements can be generated over an extended period of time, representing a large range of atmospheric conditions, surface temperatures, and viewing geometries for a relatively modest outlay. The disadvantages are uncertainties about the accuracies of the atmospheric profiles, generally derived from routine radiosonde measurements (Schmidlin, 1988), and shortcomings in the parameterization of incompletely understood physical processes in the radiative transfer model, such as the anomalous continuum absorption and emission by water vapor and the effects of tropospheric and stratospheric aerosols.

The long-term measurement of surface-emitted radiance, or the channel brightness temperatures, at the surface serves to monitor the behavior of the atmospheric correction algorithms and the performance of spaceborne radiometers. The radiance of surface-based measurements has two sources: one is of emitted radiance at the sea surface; the other is the reflected component of the downwelling radiance originating in the atmosphere. The space-based measurement is of this combination, after attenuation by atmospheric absorption and scattering, plus the radiance emitted or scattered by the atmosphere into the sensor field of view. This validation measurement is therefore less direct than a comparison of top-of-the-atmosphere data. Surface measurements can be related to space-based sensor measurements by using a radiative transfer model to estimate the atmospheric attenuation and upwelling and scattered radiation, or by converting the surface measurement to a temperature and comparing it with the surface temperature derived from the space-based measurements. In either case, successful interpretation of these data requires a good description of the atmospheric as well as the surface properties (skin SST, surface emissivity, and wind speed).

In the case where surface measurements are converted to temperature, a measurement of the downwelling radiation that is required for derivation of the temperature from the surface measurements can be achieved by pointing the surface radiometer at the sky. Suitable instruments include the M-AERI for use at sea, the AERI, and broadband infrared thermometers (Smith et al., 1996). The M-AERIs have internal blackbody calibration targets and so provide a calibrated measurement. They measure the spectrum of infrared radiation in the range of 3.3 to 18 μm with a spectral resolution of about 0.5 cm^{-1}. These spectra can be compared to the spaceborne measurements by multiplying them by the appropriate normalized channel spectral response functions. The M-AERI spectra can also be analyzed to derive surface temperature and emissivity, and, using spectra of sky radiation, the temperature and humidity structure of the atmosphere.

Broadband infrared radiation thermometers have an advantage over M-AERIs in that they are smaller and inexpensive. They usually do not have the required accuracy of 0.1 K and have only a simple internal calibration capability, if any. However, recent experience with some types indicates that they may produce useful observations and may be suitable for deployment in larger numbers on platforms of opportunity, especially if combined with a reliable external calibration assembly.

In principle, surface temperature thermometers can be deployed in numbers sufficient to provide adequate monitoring of satellite radiometer performance. However, they have a big disadvantage in that their measurement may be decoupled from remote satellite measurement by near-surface temperature gradients. For determining SST, the in situ thermometer is immersed in the water, frequently at depths of 0 to 1 m, and may register a measurement that may differ from the temperature of the radiating skin of the ocean by more than 1.0 K. These gradients are caused by heat exchange between the ocean and the atmosphere (the skin effect; see, e.g., Robinson et al., 1984; Schluessel et al., 1990) or by diurnal heating in conditions of low wind speed and therefore reduced surface mixing (see, e.g., Stramma et al., 1986). Despite this problem, in situ thermometers have been used extensively to validate satellite-measured SSTs (Strong and McClain, 1984; Llewellyn-Jones et al., 1984; Podestá et al., 1997).

DATA MANAGEMENT

Access to satellite data has been the Achilles' heel of most satellite programs. It had been argued that digital data were difficult to provide, media standards were varied, and end users did not have adequate processing capability to deal with data products. The penetration of high-performance computing into home and office settings, the explosion of Internet connectivity, and the adoption of standards for data products (Hierarchical Data Format and network Common Data Format, for example) have helped to mitigate these issues and have put pressure on data providers to build broad access to satellite products into system design. Recent and future NASA missions such as the Upper Atmosphere Research Satellite (UARS), Sea Viewing Wide Field of View Sensor (SeaWiFS), and EOS all have user-driven data system designs.

In contrast, legacy satellite systems such as NOAA AVHRR and ERS ATSR have evolved slowly and lag behind the state of the art for rapid and easy data accessibility. However, there has been a strong commitment to long-term archives from most satellite platform operators (NOAA, NASDA, ESA, and NASA). More recent systems, such as ADEOS and EOS, have built-in user-accessible browsing capabilities and data delivery via the Internet. Currently, access to operational data from the NOAA system is best provided by academic receiving and processing sites (there are a number of Internet-accessible sites in the United States). Academic sites pioneered near-real digital data access for infrared SST observations. Sites at Scripps Institution of Oceanography, the University of Miami–Rosenstiel School, and the University of Dundee have made IR SST products and data available since the late 1970s.

A major problem with SST observations and analyzed fields has been documenting system performance over the long term. NOAA now provides documentation for clock drift, sensor faults, algorithm modifications and updates, and processing system changes in a generally accessible form. NASA and NOAA have teamed up in the Pathfinder project to reprocess 15 years' worth of NOAA AVHRR SST observations with state-of-the-art algorithms to provide climate modelers and analysts a consistent set of SST fields. The magnitude of the Pathfinder effort demonstrates the need to design reprocessing capabilities for operational satellite systems. EOS has built reprocessing and maintenance of observations for validation into the Earth Observing System Data and Information System (EOSDIS).

EVOLUTION STRATEGY

When taken as the skin temperature, SST has the desirable property of being related directly to the radiance emitted from the sea surface; that is, it is a geophysical parameter. It can be sensed in both the infrared and microwave parts of the spectrum; however, the radiation temperature is dependent on the wavelength. This dependence offers opportunities to sense temperatures from several effective depths in the near-surface ocean, as well as with differing atmospheric transmissivities. Trade-offs between the infrared and microwave approaches include the accuracy and precision of the temperature estimate, needed spatial resolution, and need for all-weather observations. Historically, IR systems have had lower noise figures and better accuracy than microwave systems. However, improved microwave receivers and new reflector designs suggest that microwave-based approaches are becoming competitive with IR approaches and have the benefit of being all-weather, i.e., not limited by cloudiness. To achieve the same spatial resolution as IR sensors, however, requires very large microwave antennae.

IR approaches have also evolved. The ERS ATSR conical scan and dual blackbody calibration design has facilitated the sensing of SSTs with rms errors of 0.3 K and better. Use of slant and nadir look angles is a significant advance in IR determination of SST. It also appears to produce better aerosol corrections, resulting in an improved SST field.

These evolving approaches are limited by current knowledge of the surface processes and their diel variations, and the sparseness of skin validation observations. Current SST measurement accuracy is significantly affected by the quality and types of validation measurements (see discussion above). Diel variation in skin and bulk temperatures and related fluxes is not well understood; current research should resolve some of these issues.

REFERENCES

Abbott, M.R., and D.B. Chelton. 1991. Advances in passive remote sensing of the ocean. U.S. National Report to IUGG. Rev. Geophys. Supplement: 571-583.

Abbott, M.R., and P.M. Zion. 1987. Spatial and temporal variability of phytoplankton pigment off northern California during Coastal Ocean Dynamics Experiment 1. J. Geophys. Res. 92(C2): 1745-1755.

Barton, I.J., A.M. Zavody, D.M. O'Brien, D.R. Cutten, R.W. Saunders, and D.T. Llewellyn-Jones. 1989. Theoretical algorithms for satellite-derived sea-surface temperatures. J. Geophys. Res. 94: 3365-3375.

Bradshaw, J.T., and H.E. Fuelberg. 1993. An evaluation of HIS interferometer soundings and their use in mesoscale analysis. J. Appl. Meteorol. 32: 522-538.

Brown, O.B., and R.E. Cheney. 1983. Advances in satellite oceanography. Rev. Geophys. Space Phys. 21(5): 1216-1230.

Brown, O.B., and R.H. Evans. 1980. Interannual variability of the Somali Current system during the summer monsoon. (Abstract) EOS 61(32): 574.

Brown, O.B., J. Brown, and R. Evans. 1985. Calibration of AVHRR infrared observations. J. Geophys. Res. 90(C6): 11667-11677.

Chahine, M.T. 1980. Infrared remote sensing of sea surface temperature. In Remote Sensing of Atmospheres and Oceans, A. Deepak (ed.). New York: Academic Press.

Clancy, R.M., J. Harding, K. Pollak, and P. May. 1992. Quantification of improvements in an operational global-scale ocean thermal analysis system. J. Atmos. Ocean Technol. 9: 55-66.

Cornillon, P., and D.R. Watts. 1987. Satellite thermal infrared and inverted echo sounder determinations of the Gulf Stream northern edge. J. Atmos. Ocean Technol. 4: 712-723.

Cummings, J.A., C. Szczechowski, and M. Carnes. 1997. Global and regional ocean thermal analysis systems. Mar. Technol. Sci. J. 31(1): 63-75.

Framiñan, M.B., and O.B. Brown. 1996. Study of the Río de la Plata turbidity front, Part I: spatial and temporal distribution. Cont. Shelf Res. 16(10): 1259-1282.

Gōni, G., S. Garzoli, A. Roubicek, D. Olson, and O. Brown. 1997. Agulhas ring dynamics from TOPEX/POSEIDON satellite altimeter data. J. Mar. Res. 55: 861-883.

Harris, A.R., M.A. Saunders, J.S. Foot, K.F. Smith, and C.T. Mutlow. 1995. Improved sea surface temperature measurements from space. Geophys. Res. Lett. 22: 2159-2162.

Integrated Program Office (IPO), National Polar-orbiting Operational Environmental Satellite System (NPOESS). 1996. Integrated Operational Requirements Document (IORD) I. Joint Agency Requirements Group Administrators. 61 pp. + figures.

Jacobowitz, H., B. Chertock, S. Mango, W. Murray, and G. Ohring. 1996. National Polar-orbiting Operational Environmental Satellite System (NPOESS) Climate Requirements Workshop Report. College Park, Md.: University of Maryland.

Kamykowski, D. 1987. A preliminary biophysical model of the relationship between temperature and plant nutrients in the upper ocean. Deep-Sea Res. Oceanogr. Abstr. 34: 1067-1079.

Kelly, K.A., and S.T. Gille. 1990. Gulf Stream surface transport and statistics at 69°W from the Geosat altimeter. J. Geophys. Res. 95(C3): 3149-3161.

King, M., and D. Herring. 1992. The MODIS Airborne Simulator (MAS). The Earth Observer (November-December).

Llewellyn-Jones, D.T., P.J. Minnett, R.W. Saunders, and A.M. Zavody. 1984. Satellite multichannel infrared measurements of sea-surface temperature of the N.E. Atlantic Ocean using AVHRR/2. Q.J.R. Meteorol. Soc. 110: 613-631.

Mason, I.M., P.H. Sheather, J.A. Bowles, and G. Davies. 1996. Black body calibration sources of high accuracy for a space-borne infra-red instrument, the Along Track Scanning Radiometer. Appl. Opt. 35: 629-639.

May, D., M. Parmeter, D. Olszewski, and B. McKenzie. 1998. Operational processing of satellite SST retrievals at the Naval Oceanographic Office. Bull. Am. Meteorol. Soc. 79(3): 397-407.

McClain, E.P. 1981. Multiple atmospheric-window techniques for satellite derived sea surface temperatures. Oceanography from Space. Volume 13, J.F.R. Gower (ed.). New York: Plenum.

McClain, E.P., W.G. Pichel, C.C. Walton, Z. Ahmed, and J. Sutton. 1983. Multi-channel improvements to satellite derived global sea surface temperatures. Proc. XXIV COSPAR. Adv. Space Res. 2(6): 43-47.

McClain, E.P., W.G. Pichel, and C.C. Walton. 1985. Comparative performance of AVHRR-based multichannel sea surface temperatures. J. Geophys. Res. 90(C6): 11587-11601.

Minnett, P.J. 1986. A numerical study of the effects of anomalous North Atlantic atmospheric conditions on the infrared measurement of sea surface temperature from space. J. Geophys. Res. 91(C7): 8509-8521.

Minnett, P.J. 1990. The regional optimization of infrared measurements of sea-surface temperature from space. J. Geophys. Res. 95: 13497-13510.

Montgomery, R.S., and A.E. Strong. 1994. Coral bleaching threatens oceans, life. (Abstract) EOS 75: 145-147.

Mutlow, C.T., A.M. Zavody, I.J. Barton, and D.T. Llewellyn-Jones. 1994. Sea-surface temperature-measurements by the Along-Track Scanning Radiometer on the ERS-1 Satellite—early results. J. Geophys. Res. 99(C11): 22575-22588.

Nalli, N.R. 1995. Sea surface skin temperature retrieval using the High-Resolution Interferometer Sounder. MS Thesis. University of Wisconsin-Madison.

National Oceanic and Atmospheric Administration (NOAA). 1997. Climate Measurement Requirements for the National Polar-orbiting Operational Environmental Satellite System (NPOESS): Workshop Report, Herbert Jacobowitz (ed.), Office of Research Applications, NESDIS-NOAA, February. 77 pp.

Olson, D.B. 1991. Rings in the ocean. Annu. Rev. Earth Planet. Sci. 19: 283-311.

Podestá, G.P., and P.W. Glynn. 1997. Sea surface temperature variability in Panamá and Galápagos: extreme temperatures causing coral bleaching. J. Geophys. Res. 102: 15749-15759.

Raman, S., and D.S. Niyogi. 1998. Mesoscale analysis of a Carolina coastal front. Boundary-Layer Meteorol. 86: 125-145.

Reynolds, R., and T. Smith. 1994. Improved global sea surface temperature analysis using optimum interpolation. J. Climate 7: 929-948.

Robinson, I.S., N.C. Wells, and H. Charnock. 1984. The sea surface thermal boundary layer and its relevance to the measurement of sea surface temperature by airborne and space borne radiometers. Int. J. Remote Sensing 5: 19-46.

Rudman, S., R.W. Saunders, C.J. Kilsby, and P.J. Minnett. 1994. Water vapour continuum absorption in mid-latitudes: aircraft measurements and model comparisons. Q.J.R. Meteorol. Soc. 120: 795-807.

Saunders, R.W., and P.J. Minnett. 1990. The measurement of sea surface temperature from the C-130. MRF Internal Note No. 52. Meteorological Research Flight, Royal Aerospace Establishment, Farnborough, Hampshire, U.K. 16 pp.

Schluessel, P., H-Y. Shin, W.J. Emery, and H. Grassl. 1987. Comparison of satellite-derived sea-surface temperature with in-situ skin measurements. J. Geophys. Res. 92: 2859-2874.

Schluessel, P., W.J. Emery, H. Grassl, and T. Mammen. 1990. On the bulk-skin temperature difference and its impact on satellite remote sensing of sea surface temperatures. J. Geophys. Res. 95: 13341-13356.

Schmidlin, F.J. 1988. WMO International radiosonde intercomparison phase II, 1985. Instrument and Observing Methods, Report No. 29. WMO/TD 312. Geneva: World Meteorological Organization. 113 pp.

Shenk, W.E., and V.V. Salomonson. 1972. A multispectral technique to determine sea surface temperature using NIMBUS II data. J. Phys. Oceanogr. 2: 157-167.

Smith, A.H., R.W. Saunders, and A.M. Zavody. 1994. The validation of ATSR using aircraft radiometer data over the Tropical Atlantic. J. Atmos. Ocean Technol. 11: 789-800.

Smith, W.L., R.O. Knuteson, H.E. Revercombe, W. Feltz, H.B. Howell, W.P. Menzel, N.R. Nalli, O.B. Brown, J. Brown, P.J. Minnett, and W. McKeown. 1996. Observations of the infrared radiative properties of the ocean—implications for the measurement of sea-surface temperature via satellite remote sensing. Bull. Am. Meteorol. Soc. 77(1): 41-51.

Stramma, L., P. Cornillon, R.A. Weller, J.F. Price, and M.G. Briscoe. 1986. Large diurnal sea surface temperature variability: satellite and in situ measurements. J. Phys. Oceanogr. 16: 827-837.

Strong, A.E., and E.P. McClain. 1984. Improved ocean surface temperature from space—comparisons with drifting buoys. Bull. Am. Meteorol. Soc. 65(2): 138-142.

Susskind, J., J. Rosenfield, D. Reuter, and M.T. Chahine. 1984. Remote sensing of weather and climate parameters from HIRS2/MSU on TIROS-N. J. Geophys. Res. 89(C6): 4677-4697.

Van Scoy, K., K.P. Morris, J.E. Robertson, and A.J. Watson. 1995. Thermal-skin effect and the air-sea flux of carbon dioxide—a seasonal high-resolution estimate. Global Biogeochemical Cycles 9: 253-262.

World Ocean Circulation Experiment (WOCE). 1985. WOCE Global Air-Sea Interaction Fields, U.S. WOCE Technical Report No. 1, W.G. Largee (ed.). 36 pp.

4

Land Cover

INTRODUCTION

Land-cover and land-use changes are major components of national and international global change research programs and are topics of considerable societal relevance. Land cover—the assemblage of vegetation, exposed soil, rock, or water that occupies the land surface—and land use are significant agents of global change, with important influences on biogeochemical cycles, hydrology, and climate. Environmental change includes both anthropogenic change, caused, for example, by demographic or macroeconomic trends, and interannual, decadal, and centennial climatic trends. The environmental data records (EDRs) that specifically address vegetation assessment are the Normalized Difference Vegetation Index (NDVI; a ratio of near-infrared and red reflectance where high index values relate to the absorption of photosynthetically active radiation, a property correlated with biomass and primary production), Land Surface Temperature, Snow Cover and Depth, and Vegetation Index/Surface Type, ranked in that order by the Land Discipline Panel in the February 1997 NPOESS Climate Measurement Workshop (NOAA, 1997). Additional EDRs related to the assessment of vegetation condition and cover include Soil Moisture, Surface Albedo, and Cloud Cover. Thresholds and objectives for NPOESS land-cover EDRs are shown in Table 4.1.

BASIC SCIENCE ISSUES

Today, land-related climate-change research programs are focusing on determining the impact of land-cover and land-use change on biogeochemical cycling, on coupling land processes into global and regional climate models, and on developing an understanding of the processes that cause change (Janetos et al., 1996; Turner et al., 1995). Underlying climate-related land-cover and land-use research are fundamental questions of natural resource management and sustainable development, and the necessary integration of physical and social sciences (Turner et al., 1994).

A strong policy mandate to better understand global climate change has prioritized research on balancing the carbon budget (USGCRP, 1999). The immediate goal has been to quantify major anthropogenic greenhouse gas source terms affected by the rates of land-cover change in the tropics and the extent and frequency of fires. Understanding net terrestrial ecosystem emissions requires quantification of sinks as well as source terms, which can be done only by application of ecosystem process models. Characterizing how sources and sinks of carbon

TABLE 4.1 Planned NPOESS Land-Cover Environmental Data Records

Environmental Data Record	System Capability	Threshold	Objective
Normalized Difference Vegetation Index (NDVI)	Horizontal resolution	4 km	1 km
	Mapping accuracy	2 km	0.5 km
	Measurement range	−1 to +1	
	Measurement precision	0.04 NDVI	0.01
	Measurement accuracy	±0.05 NDVI	±0.01
	Refresh	24 h	2× per day (9:30 a.m., 1:30 p.m.)
	Long-term stability	0.04 NDVI	
Snow Cover and Depth	Sensing depth	0-50 cm	0-1 m
	Horizontal resolution	25 km	1 km
	Vertical sampling interval	>10 cm	>5, 10, 20, 30, 50, 100 cm
	Mapping accuracy	4 km	0.5 km
	Measurement accuracy	±10% clear /±20% cloudy	±30% snow depth
	Refresh	12 h	2× per day (5:30 a.m., 1:00 p.m.)
	Long-term stability	10% regional / 2% continental	5% regional / 1% continental
Land Surface Temperature	Horizontal resolution	30 km cloudy /4 km clear	12.5 cloudy /1 km clear
	Mapping accuracy	2 km	0.5 km
	Measurement range	−90 to 70 °C	
	Measurement precision	0.1 °C	0.025 °C
	Measurement accuracy	Clear ±2.8 °C	±1 °C
	Refresh	Clear: 6 h	4 h
Vegetation Index/ Surface Type	Horizontal resolution	4 km global / 4 km regional	1 km global / 0.25 km regional
	Mapping accuracy	2 km	1 km
	Measurement range	21 types	0-100% vegetation + 21 types
	Measurement accuracy	70% correct	90%
	Refresh	1× per year	4× per year
Soil Moisture[a]	Sensing depth	Thermal IR, 1 cm	Microwave, 0-5 cm
	Horizontal resolution	1 km	10 km
	Vertical sampling	1 cm	5 cm
	Mapping accuracy	0.5 km	5 km
	Measurement accuracy	±10% of total volume	±10% of total volume
	Refresh	2× per day (daytime)	Every other day

[a]For soil moisture, "Thermal IR" replaces "Threshold," and "Microwave" replaces "Objective," as explained in NOAA (1997) p. 46.

SOURCE: Extracted from NOAA (1997).

dioxide (CO_2) and other trace gases vary with land cover and land use for the major biogeochemical cycles is clearly a major research challenge. It requires a combination of satellite and airborne remote sensing, in situ measurements, process studies, and numerical modeling (Skole et al., 1997).

In general, land use is harder to quantify from space than land cover, though certain types and intensities of land use can be determined directly or indirectly. Time-series satellite data are used to provide a temporal record of changes or trends in these characteristics and the underpinning for remotely sensed land-cover research. The remotely sensed data can be used independently, combined with ancillary or in situ data, or used to parameterize or validate process models. The different model types include soil vegetation atmosphere transfer (SVAT) models, ecosystem process models, vegetation canopy structure models, land-use models, and integrated assessment models.

FUTURE DIRECTIONS

Global Biogeochemical Cycles

The recent ecological emphasis of the U.S. Global Change Research Program (USGCRP) has focused on understanding the global carbon cycle and its impacts on climate and ecosystems. Discussions of biogeochemical cycling have focused largely on its relation to atmospheric CO_2 and the potential for biological fixation of carbon. Research priorities may shift as awareness increases that human effects on the nitrogen cycle, driven primarily by use of fossil fuels and fertilizers at a rate approximating the natural biological fixation of nitrogen (Galloway et al., 1995; Howarth et al., 1996; Vitousek et al., 1997), may cause more serious environmental problems over a shorter time than the direct effects of increased atmospheric CO_2. Anthropogenically driven changes are also increasing for other biogeochemical cycles, for example, the phosphorus cycle, which may lead to potentially significant global impacts at slightly longer time horizons. These issues suggest that there will be greater scientific emphasis on observing local to regional biogeochemical impacts of global change and developing techniques for mitigation and improved land management.

It is likely that future research will focus more attention on the near-term impacts (intra-annual to decadal time periods) of climate change on regional ecosystems and how they affect the ability to provide goods and services for the growing human population, including management for control of atmospheric trace gas concentrations. Understanding the changing patterns of carbon sequestration and emissions demands better understanding of the scaling relationships between the age and structure of plant stands and regional climate processes, requiring better-resolved observations and models.

Extreme Climate Disturbances

Despite the rapid changes in land use; fragmentation of landscapes; pollution and contamination; and changes in biodiversity caused by human activities, there is little understanding about how these impacts feed back on the climate, biological, or hydrological systems. Over the next few decades the environmental consequences of anthropogenic effects may be greater than those directly attributable to increasing concentrations of atmospheric carbon.

One of the major scientific challenges to understanding and predicting the consequences of climate change on biogeochemical cycling is how to integrate land-use change given the complex interrelated temporal and spatial dynamics (NRC, 1994). These issues drive the need for improved understanding of regional climate responses, which cannot be resolved without experimental and observational studies addressing the underlying interacting time and space scales. Ecological science must develop more realistic ecosystem models to predict the consequences of changing land cover on processes. This emphasis will require direct measurements of ecosystem composition; its vertical and horizontal structures for canopy height, biomass, and surface roughness; and more detailed information on phenology—including plant litter, allocation to above- and below-ground plant components, and turnover rates.

Biodiversity

Large changes in biodiversity and loss of habitat are among the most obvious impacts of human activity on global ecosystems. Although global in extent, these impacts are expressed locally due to shifting patterns of land use, introduction of invasive nonnative species, and extinction(s) of local populations and species. If current trends continue or accelerate with growing human population density and demand for environmental goods and services, the ecosystems of the next century will become increasingly homogeneous and species richness will decline. However, it is recognized that some ecosystems may manifest more transient local heterogeneity.

The interactions between climate and land use affect the structure and function of ecosystems, which will drive changes in net primary productivity and net ecosystem productivity. Although changes in trophic food webs, such as removal of herbivores or predators, have been shown to cause substantial alterations in ecosystem structure and function (Tilman and Downing, 1994; Tilman et al., 1997; Wedin and Tilman, 1996), no ecosystem theory predicts which species, species guilds, or other structural descriptors will be essential to maintaining ecosystem function and services.

At present it is not possible to predict long-term impacts on ecosystem functioning, to know which biological components are essential for sustainability, or to develop coherent management strategies to restore, mitigate, or enhance the potential sustainability of global ecosystems. Integrated ecosystem models based on new theory are needed. Similarly integrated assessment models (IAMs) will have to be developed to predict economic and societal impacts of global change over a range of land-use practices. Models must be applicable over a range of time and space scales to improve prediction and address questions of sustainability.

CURRENT SATELLITE SAMPLING STRATEGIES

The satellite data on land use require a range of sampling strategies. Periodic sampling several times per year at high spatial resolutions (10 m to 100 m) is needed to quantify spatial change, and long-term near-daily monitoring at moderate resolutions (100 m to 4 km) is needed to characterize land cover and detect changes. Long-term measurements spanning decades are an integral part of the observation strategy for detecting and monitoring land-cover changes. Short-term observations over 1 or 2 years can be used to understand specific processes or test new technologies.

Land-cover classifications (Anderson et al., 1976) are made at high resolution, preferably using multispectral data obtained at a time optimal for discrimination. Depending on the complexity, more than one image may be used during the annual cycle to improve discrimination or in hierarchical time series (DeFries et al., 1995; Running et al., 1995). Maps of areal extent of land cover can quantify the degree of ecosystem fragmentation (Skole and Tucker, 1993). Data from successive years are used to quantify land-cover change and to characterize types of land use.

At moderate to coarse resolution, daily data can characterize vegetation phenology. These temporal vegetation patterns may be used to discriminate among vegetation types (DeFries and Townshend, 1994; Laporte et al., 1995). The requirements for global land-cover characterization have been well articulated in the literature (Townshend et al., 1994). Various approaches can be used for time-series land-cover classification. Daily data at moderate resolutions can detect change where spatial changes are large, such as burn scars from savannah or boreal fires (Justice et al., 1993; Roy et al., 1999; Kasischke and French, 1995).[1]

Very high spatial resolution data (defined as having a spatial resolution of less than 20 m) can be used to (1) characterize species composition and disturbance regimes such as insect infestations, disease, or environmental stress patterns; (2) quantify small changes in land cover; and (3) identify land-use types, such as grazing, selective logging, or mechanized farming. These data provide important information for resolving ambiguities in coarse-resolution satellite data, for addressing questions of scaling, and for monitoring carbon management. Currently such very-high-spatial-resolution satellite coverage is limited to the Russian KVR-1000 panchromatic band. The methods adopted for use with these data are derived from standard air-photo interpretation.

Securing cloud-free imagery is an overarching requirement for optical sensing systems for land-cover research. This often requires a high number of data acquisitions. Some areas are consistently cloudy and optical data are extremely difficult to obtain, sometimes even over a period of years. For such areas, microwave systems provide a marked advantage.

CURRENT OBSERVATION SYSTEMS

The Advanced Very High Resolution Radiometer (AVHRR) Global Area Coverage (GAC) multitemporal data, used extensively for land-cover classification and characterization since 1981 (Justice et al., 1985; Tucker et

---

[1]More recent literature, cited further on in the chapter, can be found, for example, in the September 1999 issue of *Photogrammetric Engineering and Remote Sensing*.

al., 1985), have been used to monitor the expansion and contraction of desert margins (Tucker et al., 1994) and changes in the length of the boreal growing season (Myneni et al., 1997a). Yet development of a consistent data set for global land-cover classification has until recently been largely inadequate (DeFries and Townshend, 1994). NASA has supported two major initiatives to develop reprocessing of the GAC data (James and Kalluri, 1994). More recently, global 1 km AVHRR data have been collected and formed into a multipurpose data set that has already been used to derive global fields of land cover (Townshend et al., 1994; Eidenshink and Faundeen, 1994; Loveland and Belward, 1997; Belward et al., 1999). Exploration into using Scanning Multichannel Microwave Radiometer (SMMR) data (25 km) demonstrated the potential use of time-series coarse-resolution microwave data for terrestrial studies (Choudhury, 1989).

A global 1 km AVHRR database was developed through the International Geosphere-Biosphere Program (IGBP) to permit improved land-cover mapping (Eidenshink and Faundeen, 1994). The same database can be used to map active fires (Justice et al., 1996). New moderate-resolution sensing systems have been built that improve on the AVHRR record for land-remote sensing (nominal resolution of 1 km), such as the European Along Track Scanning Radiometer (ATSR), Systeme Pour l'Observation de la Terre (SPOT), and SeaWiFS.

High spatial-resolution multispectral data have been available since 1972 through the Landsat program. The Landsat system has evolved technologically from the Multispectral Scanner (MSS) to the Thematic Mapper (TM). Landsat data were commercialized in the mid-1980s and data were acquired by EOSAT Corporation, based largely on customer requests. As a result, the global coverage is patchy, and data acquired by foreign ground stations are necessary to improve historical coverage. Landsat data have been the primary source for land-cover research at high spatial resolution, but the cost of commercialized Landsat data has been prohibitive for the research community. Until NASA began the Landsat Data Collection program to subsidize data purchase, researchers were not able to realize the scientific potential of these data.

With the applications potential for natural resource management and mapping, the French Centre National d'Etudes Spatiales (CNES) launched the SPOT system in the mid-1980s. A patchy historical global archive also exists for these data at SPOT IMAGE, a leading supplier of geographic information from satellites, and at various foreign ground stations. SPOT IMAGE also provides a 10-m panchromatic band to enhance its 20-m multispectral capability for land-cover mapping. More recently, the Indian Space Agency launched the Indian Remote Sensing (IRS) series of satellites, providing an additional source of high-resolution data. These data are commercially available from Space Imaging/EOSAT Corporation; however, data coverage remains incomplete.

Access to digital very-high-spatial-resolution data has traditionally been restricted to the surveillance community. Since the 1980s, digitized commercial high-resolution Russian KVR-1000 photographs (2 to 3 m) have been available for selected locations. Recent commercial initiatives have expanded the supply of declassified data to a broader community. These data appear to have considerable potential for ecosystem characterization such as community structure, tree mortality, and land-use mapping.

High-temporal-frequency data are available from geostationary satellites such as the Geostationary Operational Environmental Satellite (GOES). Such data have generally been considered too coarse in their spatial and spectral resolution for land-cover studies; however, middle-infrared subpixel fire detection has been used to sample the diurnal cycle of fire activity (Prins and Menzel, 1996). The GOES 8/9 visible and reflected infrared band is acquired at 0.9 km resolution and can be analyzed for subscenes at that resolution.

Microwave systems have been strongly supported by the European and the Japanese space agencies. The Earth Resources Satellite (ERS) and Japan's Earth Resources Satellite (JERS) high-resolution images (100 m and 12.5 m, respectively) can augment land-cover maps from optical systems. For example, JERS data are used to map inundated forests in the tropics (Saatchi et al., 1997). Multiband microwave data from the space shuttle have demonstrated the utility of synthetic aperture radar (SAR) for vegetation mapping, biomass estimation, and characterization of flooding, ice, and snow (Saatchi and Rignot, 1997; Sun and Ranson, 1998). The variable spatial resolution data from the Canadian Radarsat are now commercially available (Luscombe et al., 1993).

Aircraft data play an important role in the land-cover research program, providing a testbed for developing instruments and algorithms. Airborne simulators are available for most spaceborne sensors, including multispectral, hyperspectral, thermal infrared, radar, and microwave sensors. NASA airborne sensors test new technologies, for example, the Airborne Visible and Infrared Imaging Spectrometer (AVIRIS), a 224-band hyperspectral sensor; the

MODIS Airborne Simulator (MAS) and a multiband optical and thermal infrared sensor (MASTER); the Airborne Synthetic Aperture Radar (AirSAR), a multiband quad-polarized radar sensor; the Airborne Multi-angle Imaging Spectroradiometer (AIRMISR), a multiband, multiview-angle sensor for characterizing the radiative properties of Earth's atmosphere and land surfaces; and the Vegetation Canopy Lidar (VCL), a simulator to characterize the canopy structure.

As part of a balanced research program, NASA has invested in a number of major field campaigns aimed at improving understanding of critical land-atmosphere processes. Examples include the First ISLSCP Field Experiment (FIFE), Boreal Ecosystem-Atmosphere Study (BOREAS), Niger Hydrological-Atmospheric Pilot Experiment (HAPEX), and the Amazon Large-scale Biosphere Atmosphere (LBA) experiments, targeting specific science questions (Sellers and Hall, 1992; Sellers et al., 1996, 1997; Prince et al., 1995). Both NOAA and NASA support hydrological research under the Global Energy and Water Cycle Experiment Continental-scale International Project (GEWEX GCIP) campaign in the United States, which includes a land-surface research component (IGPO, 1994). The cost of funding comprehensive field campaigns has limited the effort largely to the examples cited above; however, such experiments are essential to scientific advances.

Ground data are also collected to validate satellite data products and provide vicarious calibration of satellite instruments (Justice et al., 1998a). A noteworthy development is the Aeronet network of Sun photometers used to characterize atmospheric optical depth and to validate atmospheric correction algorithms for multiple land-imaging satellites (Vermote et al., 1997). As recognition of the need for in situ observation networks for field testing and validation of satellite data products has increased, data collection at long-term ecological and environmental research sites is becoming integrated with the satellite data in climate-related monitoring programs (Skole et al., 1997). Examples of these sites include the national and international Long-Term Ecological Research sites, the Ameriflux and Euroflux sites (Baldocchi et al., 1996), national parks, and preserves. These sites provide a globally distributed resource for evaluating scientific data products.

Lack of data access has been an impediment to advancing climate-related land cover and land-use research. The research community has had to rely on buying low-level data products (e.g., Level 1B) from centralized archives or arranging access to higher-level products through individual principal investigators (PIs), with varying degrees of success (Justice and Townshend, 1994). The Earth Observing System (EOS) Pathfinder program marked a watershed in data availability of key AVHRR and Landsat data for the land-cover and land-use community (Maiden and Greco, 1994; Justice et al., 1995). Internationally, the IGBP data and information system (DIS) has identified some of the science requirements and coordinated the international community to help generate critical land-research data sets at the global scale, such as land cover, fire, topography, Digital Elevation Model (DEM), and gross primary productivity (GPP). Similarly, the International Satellite Land Surface Climatology Project (ISLSCP) developed a CD-ROM of multisource data sets suited to modeling land-atmosphere interactions.

OBSERVING STRATEGIES

IPO/NPOESS Plans

In the next 5 years, there will be significant improvements in the availability of satellite data for land-cover studies (Figure 4.1). The AVHRR long-term record will continue with the improved AVHRR 3 on NOAA K, L, and M, providing a global 1-km data set. The VIIRS EDR objectives for land cover appear to be based on today's AVHRR capability rather than building on the capabilities of the sensors of the coming decade. The VIIRS will replace the current AVHRR as the moderate-resolution imager.

NASA's Plans

The EOS AM platform should provide a marked improvement in moderate-resolution land remote sensing and spectral resolution (Running et al., 1994). The MODIS sensor will provide improvements over AVHRR in its spectral bandwidth, instrument calibration, signal-to-noise ratio, locational accuracy, and spatial resolution (Barnes et al., 1998). The availability of near-daily 500-m and 250-m MODIS data will lead to major advances for land-

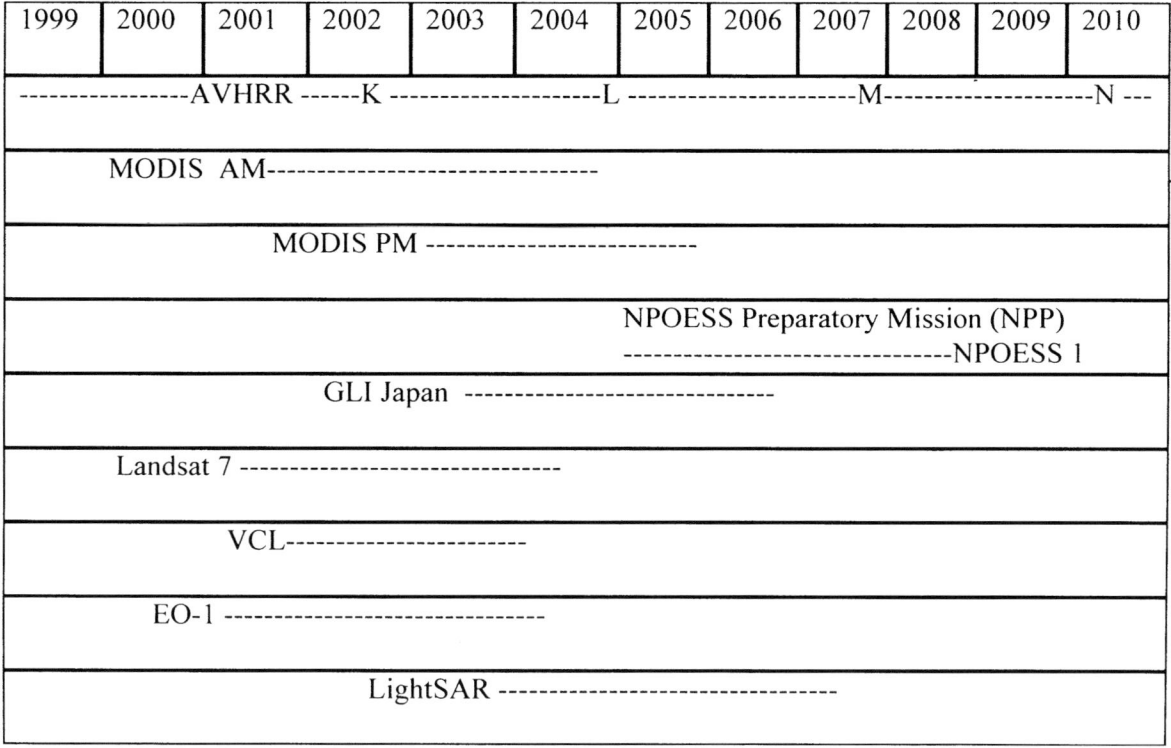

FIGURE 4.1 Time line of planned satellite observations for land-cover and land-use research. Acronyms are defined in Appendix B.

cover characterization and will mark a new era in land science and applications (Townshend and Justice, 1990). The fire detection and mapping capability of MODIS will be significantly better than that of AVHRR, which saturates at low fire temperatures, particularly after both morning and afternoon MODIS instruments are in orbit (Justice et al., 1995; Kaufman et al., 1998). New MODIS algorithms for vegetation indices, such as leaf area index/fraction of absorbed photosynthetically active radiation (LAI/FPAR), net primary productivity, bidirectional reflectance distribution function/albedo, surface temperature, snow cover, fire, land cover and land-cover change, will provide the science community a new suite of high-priority science data products (Justice et al., 1998b). The Multi-angle Imaging Spectroradiometer (MISR) could also contribute to improvements in the measurement of these canopy properties.

The ASTER instrument will provide improved multispectral optical and thermal data at high spatial resolution for land studies (Yamaguchi et al., 1998). The MISR instrument will provide new data to study surface directional reflectance properties with the potential for improved characterization of vegetation structure and atmospheric composition (Diner et al., 1998). The planned launch of the MODIS PM in 2001 will continue the MODIS data record and provide diurnal sampling and increased opportunities for cloud-free observation.

The launch of Landsat 7 and the Enhanced Thematic Mapper (ETM+) will continue the high-resolution data record and provide the next step in product continuity with technological evolution (Irons et al., 1996). In addition to better instrument performance and a new 15-m panchromatic sharpening band, Landsat 7 will for the first time have a global data acquisition strategy driven by science requirements and a major increase in data availability (Goward and Williams, 1997).

In addition to continuing and enhancing the long-term data records, some of NASA's exploratory missions target the land-cover research community. As the first of its Earth System Science Pathfinder (ESSP) missions, NASA's Vegetation Canopy Lidar (VCL) will provide improved characterization of vegetation canopy structure and offers considerable potential for detecting ecosystem disturbance, such as damage from severe storms.

LightSAR is a proposed lightweight synthetic aperture imaging radar satellite that will use advanced technologies for research, land management, and emergency response applications (NRC, 1998). LightSAR will provide all-weather, day-night, multiband, dual polarization images of most of Earth. The proposed interferometric configuration would allow development of high-resolution digital elevation maps.

The centerpiece of the Earth Observer-1 (EO-1) satellite in the New Millennium Program is the Advanced Land Imager (ALI), which will provide paired scene comparisons between the ALI and ETM+ to validate the suitability of the multispectral capability of the ALI. ALI incorporates alternative and innovative approaches to future land imaging, including a hyperspectral imaging sensor to assess the feasibility of synthesizing Landsat bands. This will open the door for advanced imaging information products for characterizing Earth's surface. Depending on the success of this mission, these technologies could provide the basis for a new generation of Landsat instruments.

International and Commercial Plans

At the international scale, plans exist for new sensors that will provide data relevant to the land-cover community, for example the Japanese Global Imager (GLI) and the European Medium Resolution Imaging Spectrometer (MERIS), to augment the moderate-resolution capability.

Plans are also under way for in situ data collection associated with the land-cover objectives. As part of its series of campaigns, NASA is investing significant resources in the Large Scale Biosphere Atmospheric Experiment in the Amazon. A smaller field campaign operating through the same period will be the Southern Africa Fires Atmosphere Research Initiative (SAFARI). In addition, a new initiative is being developed for the validation of NASA EOS data products. Plans for the EOS land-cover community have been summarized elsewhere (Justice et al., 1998b) and include a core-test-site in situ data collection program. These sites were developed from the hierarchical measurement suite proposed by the Terrestrial Observation Panel for Climate (GCOS, 1997) and provide a potential source of long-term in situ measurements. These examples are meant to be illustrative, but not encyclopedic, of the international efforts in this area.

A new multiagency national planning initiative is developing around a series of tower-based flux measurements (Baldocchi et al., 1996). The Ameriflux activity contributes to the larger international FLUXNET program aimed at producing new data sets on gas and water fluxes to provide a better understanding of primary production and land atmosphere exchanges.

INTERNATIONAL ASPECTS OF LAND-COVER OBSERVATION

Given the lack of redundancy in the planned observing systems at the national level and the plans for complementary instruments by the international community, international collaboration may be a means of reducing risk. Collaborative activities, such as the planned addition of AVHRR and possibly VIIRS instruments on the EUMETSAT METOP series, are good examples of the mutual benefits of collaboration.

International coordination on different aspects of research is continuing through international partner programs such as the IGBP International Human Dimensions Programme on Global Environmental Change (IHDP) and the World Climate Research Program (WCRP). In particular, significant progress has been made by IGBP-DIS in implementing land-related global data coordination initiatives (Townshend et al., 1994; Estes et al., 1999), the IGBP/IHDP Land-Use and Land-Cover Change (LUCC) program in identifying the leading land-cover and land-use research questions (Turner et al., 1995), IGBP-GAIM in a terrestrial model intercomparison, and the Global Energy and Water Cycle Experiment Project for Intercomparision of Land-Surface Parameterization Schemes (GEWEX PILPS) in comparing land-surface parameterizations of different general circulation models (GCMs) (Yang et al., 1995).

The International Global Observing Strategy (IGOS) initiative has provided a forum for developing the broader concepts for global observing systems and has encouraged the definition of science data requirements. As part of the Global Climate Observing System, the Terrestrial Observation Panel for Climate (TOPC) has provided the design of a hierarchical terrestrial observation network (Cihlar, 1997) and has recently started to coordinate a global net primary productivity activity. The Global Terrestrial Observing System (GTOS) has been very slow in developing, and the challenge to define the requirements for international terrestrial in situ networks remains. The Committee on Earth Observing Satellites (CEOS) provides a forum for the space agencies to coordinate their satellite programs. The CEOS Calibration/Validation working group provides a potential forum for coordinating the in situ networks needed to supplement the satellite programs. Recently CEOS has sponsored the development of pilot activities for the global observing systems. For the land-cover community, the implementation of the Global Observation of Forest Cover (GOFC) pilot study, led by the Canadian Space Agency, will have important implications for long-term land-cover monitoring for use in both science and applications.

WHAT IS NEEDED IN ADDITION TO WHAT IS PLANNED

There is a mismatch between what is needed for long-term monitoring of land cover and what is planned for NPOESS. (See Box 4.1.) In general, the land-cover EDRs (see Table 4.1) fall considerably short in specifying the long-term land data needed for climate research, especially when compared with the minimum set of terrestrial variables identified by Cihlar (1997). The land EDRs are poorly specified, and it is hard to determine the rationale for their selection and the specifications provided. The EDRs reflect a continuation of current or planned measurements rather than a set of records designed to meet climate-related research needs a decade or more from now. For the EDRs targeting the land community engaged in climate change research, the EDR objectives can be considered mandatory rather than optional, and achieving only the specified thresholds would be unacceptable.

For the currently specified land EDRs, the thresholds and objectives should be revisited. For example, the Vegetation Index/Surface Type classes should be reconsidered, because neither the 21 classes nor the associated system capabilities identified make sense in a climate measurement context. The specified levels of spatial resolution, mapping accuracy, and measurement precision need clarification or justification because, for example, the accuracy of the measurements will depend on the spectral bands chosen for the NPOESS instrument, but these are not currently specified.

A better balance is also required between continued and reliable long-term measurements and new experimental measurements. The current plans for long-term measurements could result in serious gaps and, in one case, a termination of the data record. There is little redundancy in the observation system at the national scale; a mission failure will cause major gaps in the data record. It is surprising that relatively little effort is being made by the agencies to develop an international observation strategy to alleviate this risk. Given recent experiences with mission failures, this approach appears to involve a very high risk. The committee is concerned because possible data gaps based on current plans could be filled if international cooperative plans were put in place.

A combination of programs is needed both to process and analyze data from the existing sensing systems and to support cutting-edge research that will lead to improved sensors and analysis and reduced uncertainty in climate model predictions. There should also be a clear strategy for developing tested experimental measurements into an operational suite. As part of this strategy, mechanisms should be developed and applied for moving successful technology and measurements from experimental programs into operational programs. The programs should strengthen ongoing observation programs and add observations. Current NPOESS planning appears to recognize the requirement for meeting climate data needs, but no resources are included to accommodate these needs.

Another factor to be considered in planning for new sensor missions is the uncertainty of the implications of the new national policy restricting NASA's selection of missions that may compete with private-sector investment in commercial space activities. Currently, the policy has restricted planning for hyperspectral sensors to follow the EO-1 testbed hyperspectral sensors because of a possible conflict with commercialization of the Department of Defense's (DOD's) Warfighter and Naval Earth Map Observer (NEMO) sensors. NASA has initiated discussions with DOD to evaluate the potential for these missions and technologies to meet Earth science requirements.

In addition to agreements to purchase data to fill science needs, the proposed sensors must meet data quality

BOX 4.1
Summary and Findings

Land-cover issues in the integration of the National Polar-orbiting Operational Environmental Satellite System (NPOESS) and Earth Observing System (EOS) have several implications for the National Oceanic and Atmospheric Administration (NOAA), the National Aeronautics and Space Administration (NASA), and the Integrated Program Office (IPO). There is serious concern that the NPOESS sensor will not meet the need for continuation of the long-term Advanced Very High Resolution Radiometer (AVHRR) and Landsat data records. If NPOESS is to be the continuation of long-term measurements for both the operational and science communities, the IPO should provide the planned instrument characteristics and calibration strategy for the VIIRS instrument as soon as possible. This is essential for the utility of the VIIRS data to be evaluated by the science community. For the land data products, the system capabilities must be driven by the science objectives rather than the threshold environmental data records (EDRs). Based on the current documentation, the VIIRS instrument may have to be augmented in the middle-infrared so that it can generate data products for active wildfires and burned areas at a spatial resolution sufficient to detect change.

Under current NPOESS planning, the land-cover climate research requirements for high-spatial-resolution data will not be met. This is considered a major omission for the land community engaged in climate change research. Long-term high-spatial-resolution measurements are essential for land-cover and land-use global change research. NASA may have to lead development of a high-resolution sensor and should explore a moderate-resolution imager mission (2004 to 2009) to bridge the data gap between the Moderate-resolution Imaging Spectroradiometer (MODIS) measurements and their continuation by the NPOESS VIIRS. Collocating a high-resolution imager with VIIRS is highly desirable.

In the opinion of the committee, NOAA should continue the AVHRR time series through K, L, M, and N. A 2-year overlap strategy is needed with the MODIS and NPOESS data records. There is a need to strengthen the vicarious calibration program for the red and near-infrared channels. Reprocessing of the long-term AVHRR record is needed to provide consistent multiyear data products that incorporate recent improvements in calibration and atmospheric correction and to continue the long-term data record. This could best be implemented as part of a continued NASA/NOAA data pathfinder program.

The characterization of land cover derived from the multispectral channels on Landsat TM, MODIS, and the Advanced Spaceborne Thermal Emission and Reflection radiometer (ASTER) will not be possible, or is uncertain, under the EDR descriptions. There is an urgent need for NOAA to clarify its role and commitment to the Landsat program beyond Landsat 7. For Landsat 7 and future high-resolution missions, there is also a critical need to develop real partnerships with international ground stations to augment the U.S. global acquisition strategy and secure arrangements for data provision.

In the next few years, NASA will have to demonstrate the science utility of the new EOS land instruments (MODIS, the Multi-angle Imaging Spectrometer (MISR), and ASTER) and its other near-term missions (Vegetation Canopy Lidar (VCL), EO-1, and LightSAR). This will aid in identifying new measurements that might be integrated into long-term operational systems. A mechanism will then be necessary to implement that integration. The previous Operational Satellite Improvement Program (OSIP) structure may have to be revisited. NASA should support the remote-sensing science needed to define the next generation of exploratory instruments that address the emerging key land science questions in the 2004 to 2009 period. NASA should plan to fill the gap in the long-term AM time series from the post-2004 period, prior to the launch of NPOESS (2009), and determine its role with respect to the continuation of the high-resolution record beyond Landsat 7.

standards, provide access to critical land-cover data sets, and ensure the long-term consistency of data acquisition plans and the timely availability of data. The committee is concerned that the fundamental mission requirements of these three groups—DOD, NASA, and commercial vendors—may conflict irreconcilably. The committee anticipates that the policy may lead to stopping or delaying the development of other sensors (e.g., thermal and radar) if commercial vendors have announced plans to build and launch satellites with similar or overlapping characteristics.

Although many commercial sensors have been announced (Stoney, 1997), not all of these are likely to be built, and some companies may cease operations if they are not economically viable. The fates of the Early Bird, Lewis, and Clark missions illustrate some of these problems. The implications of reliance on commercial-sector data sources for essential information needed for government programs and policy decisions require critical evaluation and development of a strategy for addressing critical data gaps.

Additional Long-Term Measurement Needs

Moderate Resolution

The VIIRS EDRs fall short of what will be needed for climate-related land-cover and land-use studies at the end of the first decade of this new century. Important data products such as LAI/FPAR, fire, net primary productivity, and land surface reflectance have been developed for near-term missions but are missing from VIIRS. The approach of providing data record requirements rather than instrument specifications to instrument designers makes it unclear whether such products could be generated once sensor packages are selected. There is no unique means of correlating instrument specifications and data record specifications; that is, there are many ways to deliver data that meet a particular requirement. There are often significant differences in other aspects of data quality that are not covered in the specifications. Scientific oversight is required at every step, not just in developing data product requirements.

With the demise of the EOS second series, the 20-year data record initially intended for the MODIS will stop with the PM mission. Assuming that sufficient characteristics of the MODIS instrument can be built into the VIIRS to meet the climate community's needs, the data record could be continued by NPOESS. However, there is likely to be a gap from the end of the MODIS AM data record (ca. 2004) to the first VIIRS launch in 2009. One highly desirable solution, in the committee's opinion, is for NASA to work with the Integrated Program Office (IPO) to develop an Advanced Global Imager for launch in the 2004 time frame which would continue many of the measurements made by MODIS and which could be extended later by VIIRS. Such a mission would provide a critical bridge between EOS and NPOESS measurements.

Through its role in the IPO, NASA is currently examining the level of augmentation with respect to engineering and instrument characterization that would be needed to accommodate its requirements. An investment by NASA to upgrade the VIIRS to permit generation of additional data products would benefit the global change science community. This would be preferable to and presumably cheaper than NASA developing its own instrument bridging MODIS and VIIRS, with no data continuity with VIIRS.

High Resolution: Optical

The lack of clear plans for long-term high-resolution optical measurements beyond Landsat 7 is perhaps the most critical gap in the land-cover and land-use part of the U.S. climate observation program. Given the increased focus on regional-scale land studies and the need for carbon monitoring, the Landsat high-resolution data record should be continued. The global change land-cover community considers continuation of this data collection essential for evaluating models and addressing scaling issues for the global data sets. The future of the Landsat program after Landsat 7 is uncertain, owing to the 1992 Title IV law requiring possible commercialization of this program or creating an international consortium for a successor sensor, and the large number of possible Landsat-like sensors. Although the Landsat program has been viewed as successful over its 25-year history and the value in maintaining the continuity of the data record is widely accepted, there is no current agreement on how to develop this next-generation sensor to meet the requirements for commercialization. For Landsat 8 to meet a 2004 launch, construction will have to begin by 2001, and these issues must be resolved by then.

A high-resolution imager could occupy space currently available on the NPOESS platform with VIIRS, creating a very attractive combination for land-cover science. Simultaneous acquisition would enable the atmospheric bands of the VIIRS to be used for atmospheric correction of the high-resolution imager, which will clearly be a requirement for a Landsat follow-on.

Experimental Missions with Associated Rationales

Hyperspectral and Very-High-Spatial-Resolution Systems

By 2000, 30 proposed satellites will be capable of providing spatial resolutions of 30 m or better (Stoney, 1997). Of these, 14 are planned commercial ventures and have pixel resolutions of 10 m or better; 8 are optical systems, and 2 are radar satellites that will be launched by other governments. The U.S. government will launch the EOS AM-1, two multispectral systems, and three satellites with hyperspectral sensors. Panchromatic data will be available at 1- to 5-m resolution from some commercial vendors. None of the four U.S. commercial satellites will have the shortwave infrared or thermal channels of Landsat TM. Four of the government satellites will have only panchromatic bands and will not be capable of multispectral imaging.

France, India, and (jointly) China and Brazil will have satellites with some Landsat-like characteristics, but because of differences in sensor characteristics and spatial resolution, it is uncertain whether they will function as replacements for Landsat. Despite the general advantages of international cooperation, the committee does have some concerns about the practicality of relying on foreign governments to supply essential satellite information, especially on a long-term basis, given that economic or national security issues could block data acquisition or distribution of data to the United States.

If all of these satellites become available, land-cover sites could be revisited at nearly 2-day intervals using data from a combination of sensors. Because of the cost of data acquisition and archiving, commercial satellite vendors have not been able to provide the types of postevent time-series data required for characterizing land-cover change. Therefore, if commercial sensors are to be the backbone in providing very-high-spatial-resolution data for the land remote-sensing community, it is essential to develop data purchase agreements that apply throughout the sensor lifetime, with the associated science acquisition strategy. Purchase agreements will have to include raw data so that calibrations can ensure consistency over time and across different sensor types.

Two NASA sensors, two proposed commercial DOD hyperspectral satellite sensors, and one commercial Australian hyperspectral sensor are planned for launch in the 2000 time frame. The proposed NASA sensors will be part of the sensor packages on NASA's EO-1 New Millennium mission; the commercially built sensors to be launched include the dual-use U.S. Air Force Warfighter I (built by Orbital Sciences Corporation for the OrbView-4 satellite), the U.S. Naval Research Laboratory's NEMO (built by Space Technology Development Corporation), and the Australian AIRES sensor. Each of these has hyperspectral capability (about 200 narrow wavelength bands) over the 400- to 2,500-nm region and high-spatial-resolution pixels (8 to 30 m). The Warfighter sensor has 80 medium wavelength infrared (MWIR) bands, which could be useful for wildfire monitoring. The commercial rationales for the DOD satellites are identification and mapping of coastal margin and ocean color patterns and land-cover spectral signature recognition. If the data are consistently available over long-term test sites, these sensors would be valuable in developing cross-sensor calibration, improving detection and understanding of seasonal to interannual changes in land cover, and providing key data to test spatial and spectral scaling algorithms.

High-Resolution Microwave

Both the optical and microwave portions of the spectrum are well suited for use in land-cover classification and estimation of forest biophysical properties. The sensor response in both spectral regions is determined to a large extent by structural attributes of the vegetation cover, differing as a consequence of sensor-specific illumination and viewing geometry. At the microscale, scattering properties are controlled by surface chemistry (i.e., pigments and water) in the optical regime and by dielectric properties (i.e., liquid water content) in the microwave regime. Hence, optical and microwave (e.g., SAR) data sets are somewhat correlated but provide much complementary information that can be exploited to improve vegetation classification and biophysical data retrievals.

Kasischke et al. (1997) reviewed the use of SAR data for land-cover classification and cited 90 percent classification accuracies for both agricultural and forested land cover. These studies commonly optimize classifications for a given airborne or orbital SAR, and comparisons (i.e., frequency and polarization configuration) are complicated by the use of diverse methodologies. Dobson et al. (2000) reported a comparison of SAR configurations for mixed-temperate forests, northern hardwoods, pine plantations, lowland forests, marshes, and prairies in northern Michigan. Shuttle Imaging Radar (SIR-C/X-SAR) polarimetric data at the L- (23 cm) and C-bands (6 cm) and vv-polarized data at the X-band (3 cm) were used to simulate SAR system scenarios representing (1) currently orbiting SAR systems (i.e., ERS-1 with C-band and vv-polarization, JERS-1 with L-band and hh-polarization, and RADARSAT with C-band and hh-polarization); (2) the combination or fusion of these systems; and (3) a suite of future orbital SAR systems under construction or in the design phase (e.g., PALSAR with dual-polarized L-band, ENVISAT with dual-polarized C-band, and LightSAR with L-band polarimetry and possible multifrequency enhancements at the C- or X-bands). A moderate-resolution SAR system capable of repeated coverage is needed as a follow-on mission to EOS for biomass retrieval, vegetation classifications, and mapping vegetation changes (Dobson et al., 1996; Kasischke et al., 1997; Sun and Ranson, 1998).

High-Temporal-Resolution Systems

Several strategies can be considered to improve the temporal resolution of land observation satellites, including use of high-resolution geostationary satellite data such as the full-spatial-resolution GOES visible and near-infrared (VNIR) image data, use of all-day and night overpasses of the various polar orbiters, or development of lightweight or low-cost constellations of polar-orbiting imaging sensors. There have been very few examples of the use of geostationary data for land studies.

CALIBRATION AND VALIDATION AND MISSION OVERLAP STRATEGIES

Both technology innovations and the demands of physically based ecosystem models require data calibrated to physical units rather than to digital numbers or radiance units. These models must account for the absorption and scattering properties of the atmosphere and for instrument characteristics and drift. Radiative transfer models must improve characterization of spatially varying atmospheric properties—specifically, aerosols, water vapor, smoke, and dust—to predict surface radiative properties more accurately. The Aeronet network of Sun photometers provides a strong foundation for the in situ component of an aerosol monitoring network, which can also be used for surface reflectance product validation.

Calibration and instrument characterization are essential for satellite-based climate measurement systems. Prelaunch instrument characterization, onboard calibration, vicarious calibration, and interinstrument cross-calibration are all critical components of a calibration program. A review of recent instrument specifications and their associated prelaunch characterization testing would help to identify areas of improvement for future systems. A closer connection between desired product characteristics and instrument specifications and tests is needed. New techniques and methods of onboard calibration that do not greatly increase weight and power requirements need to be developed and incorporated into NPOESS.

Empirical Corrections

Empirical atmospheric correction methods will continue to be integral to deriving surface reflectance and emission characteristics from satellite data. The high spatial resolution of many sensors and orientation and pointing knowledge, combined with location information of field measurements obtained from global positioning system satellites, will make it possible to calibrate sensor output accurately using vicarious methods and to validate specific pixels for specific data products. Consistent and well-registered time-series satellite and field measurements of specific, well-characterized sites are critical to developing an understanding of seasonal and interannual changes in surface conditions and evaluation of land cover. This strategy will require consistent observations, obtained repeatedly, of specific ground targets by multiple Earth-observing satellites. In addition, it requires continued acquisition of the standard in situ ecological data sets that have developed in the major field campaigns.

Selecting globally distributed land-cover sites to evaluate ecosystem data products and to establish current vegetation conditions is more difficult than using the bare-earth desert sites often selected for vicarious sensor calibration. Phenological changes and spatial complexity require a significant long-term commitment of resources at test sites. Great strides have been made in identifying and organizing global sites through the EOS Pathfinder program, the IGBP, and national and international long-term ecological programs. The primary limitation in achieving this goal is the uncertainty about institutional commitments to financially support the long-term data acquisition effort and the database infrastructure required for its use. An international initiative to coordinate test site instrumentation and data collection protocols could be envisioned in the framework of the CEOS Calibration/Validation Working Group.

Vicarious Calibration

The land-cover science community would benefit considerably from a coordinated, long-term instrument calibration program. A network of calibration sites with regular measurements could serve multiple instruments. Measurements are needed to characterize surface reflectance and emission, as well as the atmosphere over areas equivalent to several satellite footprints, and cross-calibration between in situ and satellite instruments is essential. Calibration sites are needed for both optical and thermal instruments. New initiatives are necessary to develop calibration sites suited to different parts of the spectrum. Instrumentation comparisons and ground and airborne measurement procedures should be developed to allow a distributed approach to vicarious calibration. Similarly, continuous automated vicarious methods need to be developed, building, for example, on procedures developed by Roger and Vermote (1998). Online access to regular community consensus updates on instrument calibration is essential.

Overlap and Cross-Sensor Calibrations

Since many sensors will be capable of simultaneous or near-simultaneous imaging of the same ground points, more rigorous cross-sensor calibrations and evaluation of consistency of land-cover information derived from these instruments will be obtained. This synergy may require using data products from one sensor as inputs to improve the calibration of other sensors, such as using MISR data to improve atmospheric corrections of MODIS and ASTER data. This will shift the research focus from intercalibration of data from one sensor to intercalibration of data among sensors. An overlap between successive AVHRRs and follow-on missions such as MODIS will be essential for inter-instrument cross-calibration.

Optical sensors are limited by cloud cover; therefore, developing and validating techniques to merge or fuse radar and optical data for vegetation assessment are a high priority. The high spatial resolution of LightSAR will contribute to developing this fusion. Based on theory, radar and shortwave reflected infrared bands should probe more deeply into the canopy than visible and near-infrared bands. These differences should be exploited as a basis for increasing the range of information extracted about canopy structure.

Commercial Sensors

For researchers to obtain commercially retrieved data, mechanisms such as "data buys" should be identified to allow the climate research community to incorporate commercial sensor data into the global database. This is particularly important for consistent access to very-high-spatial-resolution and hyperspectral data over calibration sites and targeted sites as these will be available primarily from commercial sensors. Raw unprocessed data must be purchased from vendors, allowing scientists to use calibration methods consistent with those applied to other sensors. Both onboard and vicarious correction methods are required for sensor calibration and validation strategies.

Ground Networks

The ground networks for nonsatellite-related data unfortunately are in decline and are not addressed by this report. The IGBP FLUXNET activity and the DOE/NASA AMERIFLUX networks provide encouraging examples of how in situ measurement systems and protocols are coordinated.

As new land data products are adopted by the modeling community, increased emphasis on product accuracy is necessary. In situ validation networks developing within the EOS program provide a strong foundation for the correlative measurement program required for product validation. The EOS land community has adopted core land-validation sites representing a range of biomes and atmospheric conditions that must be augmented and strengthened to validate future land-data products. Coordinated and periodic ground-based measurement of such variables as surface reflectance, LAI/FPAR, canopy structure land cover, fire-burned area, and net primary productivity will provide a critical component for assessing product accuracy. A research challenge is to tie measurements to the equivalent satellite products. In particular, continuous tower-based (point) eddy flux measurements will have to be related to satellite and model-derived net primary productivity.

Future systems generating higher-order data products such as the VIIRS should include an explicit validation component to enable science users to characterize error budgets in their analysis. Automated and reliable in situ instrumentation is highly desirable. The validation community should continue to develop protocols for collecting data for validation and for instrument calibration that allow a distributed and international validation system to be developed.

DATA PROCESSING AND MANAGEMENT

Data processing, archiving, and distribution are important components of a land-cover monitoring system. Characterization and monitoring of data quality are essential to the data production chain. Procedures for operational quality assurance associated with land data production should be developed and applied to future data systems.

The federation concept for data archiving is being prototyped over the next three years through the Earth System Science Information Partnerships (ESSIPS), which will provide additional and alternative services to those of the EOSDIS Core System (ECS). The ESSIPS will move the community toward a more distributed data system. As part of its current approach, NASA is also involving scientists more closely in the data system design and implementation through what it calls "PI processing," ensuring greater ownership of the data system and products by the science community. Successive reprocessing should be built into the plans for data management to ensure a consistent data record.

Processing of satellite data for land-cover analysis often involves using ancillary products such as digital elevation models and assimilated climate data. Timely provision of these data is an important part of the data chain. Arrangements have to be made for provision and archiving of these data sets for initial processing, and also for reprocessing of these data.

THE NECESSARY OBSERVATION STRATEGY

A Multisatellite, Multisensor Approach

Current methods typically involve analyzing single-sensor data, either as a one-time image or as a multitemporal data set. Limited access to radar and thermal sensor data has meant that there is little history of combining data from different spectral regions to optimize characterization of land-cover properties. The availability of new sensors will likely place greater emphasis on fusing data using (1) multiple sensors for multiresolution analysis and statistical subsampling approaches or (2) sensor suites, that is, combinations of optical and radar or optical and thermal sensors, that provide collocated and coincident temporal data. Better understanding of the spectral, spatial, and temporal scaling issues and use of high-resolution data as subsampling for moderate-resolution sensors is necessary.

The spatial scale of regional land-cover problems requires higher-resolution data than the 1 to 4 km for VIIRS to characterize surface conditions and quantify the rates of land-cover change. A higher spatial scale is essential to understanding global change mechanisms, while scaling studies may reduce uncertainties in interpretation of VIIRS data. Within the next 5 years, national computing capacity is likely to increase to make land parameterization at a 1 km grid feasible. This will drive a demand for collection of moderate-resolution data, e.g., 250 to 500 m.

Improved and New Measurements

One area where improved resolution is necessary is in the estimation of LAI/FPAR, which will lead to significant improvements in estimates of net primary productivity at midlatitudes and high latitudes. The current emphasis on the use of NDVI has been largely driven by spectral availability of long-term data sets. The limitations of NDVI in certain habitats (very high or very low cover) has resulted in the development of a number of derived indices to improve this product, but no method has been shown to perform better consistently. Improving estimates of vegetation phenology will reduce errors in net primary productivity and net ecosystem productivity. Direct measurement of biomass and vertical and horizontal vegetation structure is also needed to parameterize and validate ecosystem models.

NASA EOS sensors may produce new information on canopy water content, separation of photosynthetic pigments, dry plant matter, and soil quality. Canopy water content increases linearly with LAI, greatly increasing the range of sensitivity derived from NDVI measurements (Roberts et al., 1997; Asner et al., 1998). Methods using radiative transfer appear promising (Myneni et al., 1997b). Separation of photosynthetic pigments (Penuelas et al., 1995; Gamon et al., 1997) could eliminate much of the error in modeled net primary productivity estimates and significantly reduce uncertainties in biogeochemical budgets and estimates of evapotranspiration. Improved estimates of dry plant matter greatly reduce errors in estimates of surface albedo, soil processes, erosion, and wildfire hazard. This information would improve wildfire assessments (Ustin et al., 1998; Roberts et al., 1998) and estimates of biomass and decomposition.

Monitoring land-cover change, the type and intensity of changes, and evidence of land degradation will benefit from new vegetation indices using other spectral regions. Change-detection accuracy will improve with the use of calibrated, atmospherically corrected surface reflectance data. Precise multitemporal registration of time-series data is essential. Issues related to understanding biodiversity changes, species invasions, and the changing distribution of tree ages in forests, from older to younger stands, require new types of data analysis. Both LightSAR and VCL promise to contribute to this assessment, as do methods derived from hyperspectral and very-high-spatial-resolution sensors. New methods that include use of neural nets, wavelets, mixture and multiple end-member models, and inverse canopy modeling developed for high-spatial-resolution mapping of vegetation types need more critical evaluation over a wide variety of global biomes.

While changes in large patterns of land cover resulting from land use and wildfires will be identifiable using VIIRS data specified at its current land-cover EDR objective of 250 m, it is certain that higher-spatial-resolution sensors (about 20 m) will also be needed to provide direct quantification of land use, types, and intensity of use and land degradation. The VIIRS EDR for land classification will be inadequate for this assessment.

AREAS FOR RESEARCH AND DEVELOPMENT

Among the areas associated with satellite observations that are appropriate for further research and development by the land-cover community are the following examples:

- Development of multiscale sampling strategies for forest monitoring, such as combining MODIS and Landsat 7;
- Development of automated change detection methodologies;

- Evaluation of the role of very-high-spatial-resolution and hyperspectral data for land-cover and land-use characterization;
- Development of approaches to monitoring and modeling changes in biodiversity at the landscape scale;
- Development of product validation protocols for addressing issues of surface heterogeneity, such as scaling from point-flux measurements to satellite footprints;
- Development of methods for data fusion that will allow analysis of data from multiple sensing systems at multiple resolutions, including SAR and VCL; and
- Development of tools for managing large volumes of data, desktop data processing, and automated data retrieval and analysis.

BIBLIOGRAPHY

Anderson, J.R., E.E. Hardy, J.T. Roach, and R.E. Witner. 1976. A land use land cover classification system for use with remote sensing data. Geological Survey Professional Paper 964. Washington, D.C.: U.S. Government Printing Office.

Asner, G.P., C.A. Wessman, D.S. Schimel, and S. Archer. 1998. Variability in leaf and litter optical properties: implications for BRDF model inversions using AVHRR, MODIS, and MISR. Remote Sensing Environ. 63(3): 243-257.

Baldocchi, D., R. Valentini, S. Running, W. Oechel, and R. Dahlman. 1996. Strategies for measuring and modelling carbon dioxide and water vapor fluxes over terrestrial ecosystems. Global Change Biology 2(3): 159-168.

Barnes, W.L., T.S. Pagano, and V.V. Salomonson. 1998. Prelaunch characteristics of the Moderate Resolution Imaging Spectroradiometer (MODIS) on EOS-AM. IEEE Trans. Geosci. Remote Sensing 36(4): 1088-1100.

Belward, A., J. Estes, T., Loveland. 1999. The IGBP-DIS Global 1-km Land Cover Data Set DISCover: a Project Overview. Photogrammetric Engineering and Remote Sensing 65(9): 1013-1020.

Choudhury, B.J. 1989. Monitoring global land surface using Nimbus-7 37 GHz polarization difference. Int. J. Remote Sensing 10: 1579-1605.

Cihlar, J. (ed.). 1997. GCOS/GTOS Plan for Terrestrial Climate Related Observations. Version 2. GTOS Report 11, WMO TD No. 796, UNEP-DEIA-Tr 97-7. 125 pp.

DeFries, R.S., and J.R.G. Townshend. 1994. NDVI-derived land cover classification at global scales. Int. J. Remote Sensing 15(17): 3567-3586.

DeFries, R.S., C.B. Field, F. Inez, C.O. Justice, P.A. Matson, M. Matthews, H.A. Mooney, C.S. Potter, K. Prentice, P.J. Sellers, J. Townshend, C.J. Tucker, S.L. Ustin, and P.M. Vitousek. 1995. Mapping the land surface for global atmosphere-biosphere models: toward continuous distributions of vegetation's functional properties. J. Geophys. Res. 100(D10, 20): 867-882.

Delcourt, H.R., and P.A. Delcourt. 1996. Presettlement landscape heterogeneity: evaluating grain of resolution using General Land Office Survey data. Landsc. Ecol., 11(6): 363-381.

Diner, D.J., J.C. Beckert, T.H. Reilly, C.J. Bruegge, J.E. Conel, R.A. Kahn, J.V. Martonchik, T.P. Ackerman, R. Davies, S.A.W. Gerstl, H.R. Gordon, J-P. Muller, R.B. Myneni, P.J. Sellers, B. Pinty, and M.M. Vertraete. 1998. Multi-angle imaging spectroradiometer (MISR) instrument description and experiment overview. IEEE Trans. Geosci. Remote Sensing 36(4): 1072-1087.

Dobson, M.C., L.E. Pierce, and F.T. Ulaby. 1996. Knowledge-based land-cover classification using ERS-1/JERS-1 SAR composites. IEEE Trans. Geosci. Remote Sensing 34(1): 83-99.

Dobson, M.C., L.E. Pierce, and F.T. Ulaby. 2000. Evaluation of SAR sensor configurations for terrain classification and forest biophysical retrievals. Int. J. Remote Sensing, in press.

Eidenshink, J.C., and J.L. Faundeen. 1994. The 1 km AVHRR global land data set: first stages in implementation. Int. J. Remote Sensing 15(17): 3443-3462.

Estes, J., A. Belward, T. Loveland, J. Scepan, A. Strahler, J.R.G. Townshend, and C.O. Justice. 1999. The Way Forward. Photogrammetric Engineering and Remote Sensing 65(9): 1089-1093.

Galloway, J.N., W.H. Schlesinger, H. Levy, A. Michaels, and J.L. Schnoor. 1995. Nitrogen fixation—anthropogenic enhancement-environmental response. Global Biogeochemical Cycles 9(2): 235-252.

Gamon, J.A., L. Serrano, and J.S. Surfus. 1997. The photochemical reflectance index: an optical indicator of photosynthetic radiation use efficiency across species, functional types, and nutrient levels. Oecol. Plant. 112(4): 492-501.

Global Climate Observing System (GCOS). 1997. Global Hierarchical Observing Strategy (GHOST). GCOS Report 33, WMO TD No. 862. World Meteorological Organization. June 1997.

Goward, S.N., and D.L. Williams. 1997. Landsat and Earth systems science: development of terrestrial monitoring. Photogramm. Eng. Remote Sensing 63(7): 887-900.

Hansen, A.J., S.L. Garman, J.F. Weigand, D.L. Urban, et al. 1995. Alternative silvicultural regimes in the Pacific Northwest—simulations of ecological and economic effects. Ecol. Appl. 5(3): 535-554.

Harrell, P.A., E.S. Kasischke, L.L. Bourgeau-Chavez, E.M. Haney, and N.L. Christensen, Jr. 1997. Evaluation of approaches to estimating aboveground biomass in southern pine forests using SIR-C data. Remote Sensing Environ. 59(2): 223-233.

Howarth, R.W., G. Billen, D. Swaney, A. Townsend, et al. 1996. Regional nitrogen budgets and riverine N-and-P fluxes for the drainages to the North Atlantic Ocean—natural and human influences. Biogeochemistry 35(1): 75-139.

Hussin, Y.A., R.M. Reich, and R.M. Hoffer. 1991. Estimating slash pine biomass using radar backscatter. IEEE Trans. Geosci. Remote Sensing 29(3): 427-431.

International GEWEX Project Office (IGPO). 1994. Implementation Plan for the GEWEX Continental Scale International Project (GCIP): Volume II, Research. Washington, D.C.: International GEWEX Office.

Irons, J.R., D.L. Williams, and B.L. Markham. 1996. Landsat 7 and Beyond. IGARSS'96, May 20-24, 1996, Lincoln, Nebr.: 2161-2163.

James, M.E., and S.N.V. Kalluri. 1994. The Pathfinder AVHRR land data set: an improved coarse resolution data set for terrestrial monitoring. Int. J. Remote Sensing 15(17): 3347-3364.

Janetos, A.C., C.O. Justice, and R.C. Harriss. 1996. Mission to Planet Earth: land cover and land use change. Biomass Burning and Global Change, J.S. Levine (ed.). Cambridge, Mass.: Massachusetts Institute of Technology.

Justice, C.O., J.R.G. Townshend, B.N. Holben, and C.J. Tucker. 1985. Analysis of the phenology of global vegetation using meteorological satellite data. Int. J. Remote Sensing 6(8): 1272-1318.

Justice, C.O., J.P. Malingreau, and A. Setzer. 1993. Satellite remote sensing of fires: potential and limitation. Fire In the Environment; Its Ecological, Climatic and Atmospheric Chemical Importance, P. Crutzen and J. Goldammer (eds.). Chichester: John Wiley and Sons.

Justice, C.O., and J.R.G. Townshend. 1994. Data sets for global remote sensing: lessons learnt. Int. J. Remote Sensing 15(17): 3621-3639.

Justice, C.O., J.D. Kendall, P.R. Dowty, and R.J. Scholes. 1996. Satellite remote sensing of fires during the SAFARI Campaign using NOAA-AVHRR data. J. Geophys. Res. (101): 23851-23863.

Justice, C.O., D. Starr, D. Wickland, J. Privette, and T. Suttles. 1998a. EOS validation coordination: an update. Earth Observer 10(3): 55-60.

Justice, C.O., E. Vermote, J.R.G. Townshend, R. DeFries, D.R. Roy, D.K. Hall, V.V. Salomonson, J.L. Privette, G. Riggs, A. Strahler, W. Lucht, R. Myneni, Y. Knyazikhin, S. W. Running, R.R. Nemani, Z. Wan, A. Huete, W. van Leeuwen, R.E. Wolfe, L. Giglio, J-P. Muller, P. Lewis, and M.J. Barnsley. 1998b. The Moderate Resolution Imaging Spectroradiometer (MODIS): land remote sensing for global change research. IEEE Trans. Geosci. Remote Sensing 36(4): 1228-1249.

Justice, C.O., G.B. Bailey, M.E. Maiden, S.J. Rasool, et al. 1995. Recent data and information system initiatives for remotely sensed measurements of the land surface. Remote Sensing Environ. 51(1): 235-244.

Kasischke, E.S., and N.H.F. French. 1995. Locating and estimating the areal extent of wildfires in Alaskan boreal forests using multiple-season AVHRR NDVI composite data. Remote Sensing Environ. 51: 263-275.

Kasischke, E.S., J.M. Melack, and M.C. Dobson. 1997. The use of imaging radars for ecological applications—a review. Remote Sensing Environ. 59(2): 141-156.

Kaufman, Y.J., C.O. Justice, L. Flynn, J. Kendall, E. Prins, D.E. Ward, P. Menzel, and A. Setzer. 1998. Potential global fire monitoring from EOS-MODIS. J. Geophys. Res. 103, D24, 32, 215-32, 238.

Kellndorfer, J.M., L.E. Pierce, M.C. Dobson, and F.T. Ulaby. 1998. Toward consistent regional-to-global-scale vegetation characterization using orbital SAR systems. IEEE Trans. Geosci. Remote Sensing 36(N5 PT1): 1396-1411.

Kimes, D.S., K.J.Ranson, and G. Sun. 1997. Inversion of a forest backscatter model using neural networks. Int. J. Remote Sensing 18(10): 2181-2199.

Laporte, N., C.O. Justice, and J. Kendall. 1995. Mapping the humid forests of Central Africa using NOAA–AVHRR data. Int. J. Remote Sensing 16(6): 1127-1145.

Los, S.O., C.O. Justice, and C.J. Tucker. 1994. A global 1-degrees-by-1-degrees NDVI data set for climate studies derived from the GIMMS continental NDVI data. Int. J. Remote Sensing 15(17): 3493-3518.

Loveland, T.R., and A.S. Belward. 1997. The IGBP-DIS Global 1 km Land Cover Data Set, DISCover First Results. Int. J. Remote Sensing 18(15): 3289-3295.

Luscombe, A.P., I. Ferguson, N. Shepherd, D.G. Zimcik, and P. Naraine. 1993. The RADARSAT synthetic aperture radar development. Can. J. Remote Sensing 19(4): 298-310.

Maiden, M.E., and S. Greco. 1994. NASA's Pathfinder data set programme: land surface parameters. Int. J. Remote Sensing 15(17): 3333-3345.

Martens, S.N., S.L. Ustin, and R.A. Rousseau. 1993. Estimation of tree canopy leaf area index by gap fraction analysis. For. Ecol. Manage. 61: 91-108.

Myneni, R.B., C.D. Keeling, C.J.Tucker, G.Asrar, and R.R. Nemani. 1997a. Increased plant growth in the northern latitudes from 1981 to 1991. Nature 386: 698-702.

Myneni, R.B., R.R. Nemani, and S.W. Running. 1997b. Estimation of global leaf area index and absorbed PAR-photosynthetically active radiation using radiative transfer models. IEEE Trans. Geosci. Remote Sensing 35(6): 1380-1393.

National Oceanic and Atmospheric Administration (NOAA). 1997. Climate Measurement Requirements for the National Polar-orbiting Operational Environmental Satellite System (NPOESS), Workshop Report, Herbert Jacobowitz (ed.), Office of Research Applications, NESDIS-NOAA, February. 77 pp.

National Research Council (NRC). 1994. The Role of Terrestrial Ecosystems in Global Change: A Plan for Action. Washington, D.C.: National Academy Press.

National Research Council (NRC). 1998. Development and Application of Synthetic Small Spaceborne Aperture Radars. Washington, D.C.: National Academy Press.

Penuelas, J., I. Filella., and J.A. Gamon. 1995. Assessment of photosynthetic radiation-use efficiency with spectral reflectance. New Phytol. 131(3): 291-296.

Prince, S.D., Y.H. Kerr, J.P. Goutorbe, T. Lebel, A. Tinga, P. Bessemoulin, J. Brouwer, A.J. Dolman, E.T. Engman, J.H.C. Gash, M. Hoepffner, P. Kabat, B. Monteny, F. Said, P. Sellers, and J. Wallace. 1995. Geographical, biological and remote sensing aspects of the Hydrologic Atmospheric Pilot Experiment in the Sahel (HAPEX-Sahel). Remote Sensing Environ. 51: 215-234.

Prins, E.M., and W.P. Menzel. 1996. Investigation of biomass burning and aerosol loading and transport utilizing geostationary satellite data. Pp. 65-72 in Biomass Burning and Global Change, J.S. Levine (ed.). Cambridge, Mass.: MIT Press.

Remer, L.A., S. Gasso, D.A. Hegg, Y.J. Kaufman, et al. 1997. Urban/industrial aerosol: Ground-based Sun/sky radiometer and airborne in situ measurements. J. Geophys. Res. Atmos. 102(D14): 16849-16859.

Roberts, D.A., R.O. Green, and J.B. Adams. 1997. Temporal and spatial patterns in vegetation and atmospheric properties from AVIRIS. Remote Sensing Environ. 62(3): 223-240.

Roberts, D.A., M. Gardner, R. Church, S.L. Ustin, G. Scheer, and R.O. Green. 1998. Mapping chaparral in the Santa Monica Mountains using multiple endmember spectral mixture models. Remote Sensing Environ. 65: 267-279.

Roger, J.C., and E.F. Vermote. 1998. A method to retrieve the reflectivity signature at 3.75 mu from AVHRR data. Remote Sensing Environ. 64(1): 103-114.

Roy, D.P., L. Giglio, J. Kendall, and C.O. Justice. 1999. Multitemporal active-fire based burn scar detection algorithm. Int. J. Remote Sensing 20: 1031-1038.

Running, S.W., T.R. Loveland, L.L. Pierce, R.R. Nemani, and E.R. Hunt. 1995. A remote sensing based vegetation classification logic for global land cover analysis. Remote Sensing Environ. 51: 39-48.

Running, S.W., C.O. Justice, V.V. Salomonson, D. Hall, J. Barker, Y.J. Kaufman, A.R. Strahler, J-P. Muller, V. Vanderbilt, Z.M. Wan, P. Teillet, and D. Carneggie. 1994. Terrestrial remote sensing science and algorithms planned for the MODIS-EOS. Int. J. Remote Sensing 15(17): 3587-3620.

Saatchi, S., J. Soares, and D. Alves. 1997. Mapping deforestation and land use in Amazon rainforest by using SIR-C imagery. Remote Sensing Environ. 59(2): 191-202.

Saatchi, S., and E. Rignot. 1997. Classification of boreal forest cover types using SAR images. Remote Sensing Environ. 60(3): 270-281.

Sellers, P.J., and F.G. Hall. 1992. FIFE in 1992: Results, scientific gains and future directions. J. Geophys. Res. 97: 19091-19109.

Sellers, P.J., F.G. Hall, R.D. Kelly, A. Black, D. Baldocchi, J. Berry, M. Ryan, K.J. Ranson, P.M. Crill, D.P. Lettenmaier, H. Margolis, J. Cihlar, J. Newcomer, D. Fitzjarrald, P.G. Jarvis, S.T. Gower, D. Halliwell, D. Williams, B. Goodison, D.E. Wickland, and F.E. Guertin. 1997. BOREAS in 1997: Experiment overview, scientific results, and future directions. J. Geophys. Res. 102(D24): 28731-28769.

Sellers, P.J., B.W. Meeson, J.Closs, J. Collatz, et al. 1996. The ISLSCP initiative I global datasets—surface boundary conditions and atmospheric forcings for land-atmosphere studies. Bull. Am. Meteorol. Soc. 77(9): 1987-2005.

Shugart, H.H. 1998. Terrestrial ecosystems in changing environments. Cambridge Studies in Terrestrial Ecology. United Kingdom: Cambridge University Press.

Skole, D.L., and C.J. Tucker. 1993. Tropical deforestation, fragmented habitat and adversely affected habitat in the Brazilian Amazon: 1978-1988. Science 260: 1905-1910.

Skole, D.L., C.O. Justice, A. Janetos, and J.R.G. Townshend. 1997. A Land Cover Change Monitoring Program: A Strategy for International Effort. Mitigation and Adaptation Strategies for Global Change, 2, 2-3. The Netherlands: Kluwer Academic Publishers.

Stoney, W.E. 1997. Land Sensing Satellites in the Year 2000. International Geoscience and Remote Sensing Symposium (IGARSS), Singapore, August 7. Available online at <http://geo.arc.nasa.gov/sge/landsat/wesigrs.html>.

Sun, G., and K.J. Ranson. 1998. Radar modelling of forest spatial patterns. Int. J. Remote Sensing 19(9): 1769-1791.

Tilman, D., J. Knops, D.Wedin, P. Reich, et al. 1997. The influence of functional diversity and composition on ecosystem processes. Science 277(5330): 1300-1302.

Tilman, D., and J.A. Downing. 1994. Biodiversity and stability in grasslands. Nature 367(6461): 363-365.

Townshend, J.R.G., and C.O. Justice. 1990. The spatial variation of vegetation changes at very coarse scales. Int. J. Remote Sensing 11(1): 149-157.

Townshend, J.R.G., C.O. Justice, D. Skole, J.P. Malingreau, J. Cihlar, P. Teilliet, F. Sadowski, and S. Ruttenberg. 1994. The 1 km resolution global data set: Needs of the International Geosphere Biosphere Program. Int. J. Remote Sensing 15(17): 3417-3442.

Tucker, C.J., W.W. Newcomb, and H.E. Dregne. 1994. AVHRR data sets for determination of desert spatial extent. Int. J. Remote Sensing 15(17): 3547-3565.

Tucker C.J., J.R.G. Townshend, and T.E. Goff. 1985. African land cover classification using satellite data. Science 227: 369-375.

Turner B.L., W.B. Meyer, and D.L. Skole. 1994. Global land use/land cover change: towards an integrated program of study. Ambio 23(1): 91-95.

Turner B.L., D. Skole, S. Sanderson, G. Fischer, L. Fresco, and R. Leemans. 1995. Land-use and land cover change: Science Research Plan. IGBP Report 35. Stockholm: IGBP. 123 pp.

U.S. Global Change Research Program (USGCRP). 1999. Our Changing Planet: The FY 2000 U.S. Global Change Research Program. U.S. Global Change Research Program Office. Washington, D.C., March.

Ustin, S.L., D. Roberts, S. Jacquemoud, M. Gardner, G. Scheer, C.M. Castaneda, and A. Palacios-Orueta. 1998. Estimating canopy water content of chaparral shrubs using optical methods. Remote Sensing Environ. 65: 280-291.

Vermote E.F., N. El Saleous, C.O. Justice, Y.K. Kaufman, J.L. Privette, L. Remer, J.C. Roger, and D. Tanre. 1997. Atmospheric correction of visible to middle infrared EOS MODIS data over land surfaces: background, operational algorithm and validation. J. Geophys. Res. 102: 17131-17141.

Vitousek, P.M., J.D. Aber, R.W. Howarth, G.E. Likens, P.A. Matson, D.W. Schindler, W.H. Schlesinger, and D.G. Tilman. 1997. Human alteration of the global nitrogen cycle: sources and consequences. Ecol. Appl. 7(3): 737-750.

Wedin, D.A., and D. Tilman. 1996. Influence of nitrogen loading and species composition on the carbon balance of grasslands. Science 274(5293): 1720-1723.

Xie, H., L.E. Pierce, M.C. Dobson, and F.T. Ulaby. 1998. Combining Orbital SAR and Optical Data for Global Classification, IGARSS'98, July 6-10, 1998, Seattle, Wash.

Yamaguchi, Y., A.B. Kahle, H. Tsu, T. Kawakami, and M. Pniel. 1998. Overview of the Advanced Spaceborne Thermal Emission and Reflection Radiometer (ASTER). IEEE Trans. Geosci. Remote Sensing 36(4): 1062-1071.

Yang, Z-L., R.E. Dickinson, A. Henderson-Sellers, and A.J. Pitman. 1995. Preliminary study of spin up processes in land surface models with the first stage of the Project for Intercomparison of Land Surface Parameterization Schemes Phase 1(a). J. Geophys. Res. Atmos. 100(16): 16533-16578.

Zavody, A.M., C.T. Mutlow, and D.T. Llewellyn-Jones. 1995. A radiative transfer model for sea surface temperature retrieval for the Along-Track Scanning Radiometer. J. Geophys. Res. 100: 937-952.

5

Ocean Color

INTRODUCTION

Marine phytoplankton account for nearly half of the world's total primary productivity. They dominate the structure of the upper trophic levels of the food web and play a critical role in the cycling of biogeochemical properties. Biomass turnover rates for plankton ecosystems are 100 times faster than those for terrestrial systems, leading to a close relationship between upper-ocean ecology and physical forcing. For example, coupled ocean and atmosphere models show that changes in the phytoplankton community structure and the resulting elemental interactions can drastically affect the rate of carbon dioxide (CO_2) increase in the atmosphere. Moreover, the large space and time scales associated with ocean biogeochemistry and circulation can be disrupted on far shorter time scales, such as those of El Niño/Southern Oscillation (ENSO) events. This coupling of large and small scales leads to the fundamental sampling requirement of global-scale, long time series (decades) at moderate time and spatial scales (days and kilometers).

BASIC SCIENCE ISSUES

In addition to ocean biogeochemistry and its linkage with climate, the productivity of the ocean is the fundamental limit on the number of fish that can be harvested. Because of the near-collapse of many of the world's fisheries, there is renewed interest in understanding the complex linkage between primary productivity and fisheries production. As with ocean biogeochemistry, the scales are both large and small. However, because coastal zones are often used as nurseries, there is particular interest in the smaller scales associated with the nearshore environments.

Another objective focuses on the coastal zone itself as a buffer between human activities on the land and the deep ocean. In addition to physical and biological coupling, there is particular interest in the impacts of coastal pollution, either through direct discharge into the sea or by riverine inputs. Harmful algal blooms (e.g., red tides) can have drastic economic and human health impacts. Given the economic and recreational importance of the coastal zone, there is increasing demand for high-resolution coastal ocean monitoring and prediction.

The science issues can thus be summarized as three broad objectives:

1. Predict the ocean's biogeochemical response to and its influence on climate change;
2. Predict the variability in the structure of the phytoplankton community and its links with higher trophic levels as well as ocean biogeochemistry; and
3. Develop the scientific basis necessary to manage the sustainable resources of the coastal marine ecosystem effectively.

Current Approaches in Remote Sensing

To address these three objectives, the remote sensing community has developed an evolving suite of satellite sensors to collect measurements of ocean color in the visible portion of the electromagnetic spectrum. The basic measurement of phytoplankton biomass relies on the strong absorption of visible light by chlorophyll (the primary light-harvesting pigment), which peaks near 443 nm (Gordon and Morel, 1983; Kirk, 1994). This absorption characteristic is a robust feature across a broad range of productivity levels in the world's oceans.

Atmospheric Correction

The challenge for spaceborne sensors is that 80 to 90 percent of a satellite-sensed signal originates in the atmosphere (Gordon and Morel, 1983). Much of this atmospheric signal is Rayleigh (or molecular) scattering, primarily from stratospheric ozone. After accounting for satellite and solar geometry for a particular scene, it is relatively straightforward to make corrections based on knowledge (or estimates) of extraterrestrial solar radiation, ozone concentration, and atmospheric pressure. However, aerosol scattering, primarily from hydrophilic particles in the marine boundary layer, is a much more complex process. It varies strongly as a function of time and location. Because it is not yet possible to make direct measurements of these aerosols and their contribution to atmospheric optical properties, the remote sensing community has relied on an indirect approach. Because the ocean is largely "black" in the red and near-infrared portion of the spectrum, it can be assumed that any radiance measured at these wavelengths originated in the atmosphere and was not backscattered out of the ocean. Relying on ratios between the remaining wavelengths, the spectral dependence of aerosol scattering can be propagated down to the short wavelengths in the blue portion of the visible light spectrum.

The atmospheric correction schemes have matured considerably over the past 20 years since the launch of the first ocean color sensor, the Coastal Zone Color Scanner (CZCS), on Nimbus-7. The original atmospheric schemes relied on locating a "clear-water pixel" where chlorophyll concentrations were low and the spectrum of water-leaving radiance therefore well known. The atmospheric correction for this clear-water portion of the image was then extrapolated across the entire scene. Obviously there are serious limitations to this approach; for example, there may not be a low-chlorophyll pixel in the image, or atmospheric properties may change significantly within an image that covers nearly a million square kilometers. The next step was to enable pixel-by-pixel correction, thus eliminating the need for an imagewide correction and a low-chlorophyll region.

As analysis and processing of CZCS data continued, it became apparent that the atmospheric correction schemes had to accommodate multiple scattering by molecules. The first-generation algorithms assumed that a photon would be scattered only once. However, at low Sun angles or at the edge of the sensor swath, the probability of multiple Rayleigh scattering increased substantially. Moreover, post-CZCS sensors—e.g., the Sea-viewing Wide Field of View Sensor (SeaWiFS)—have substantially higher signal-to-noise ratios (SNRs), which means processes such as Rayleigh-aerosol scattering become significant. Atmospheric correction algorithms for the Moderate-resolution Imaging Spectroradiometer (MODIS) incorporate an approach to these issues, and researchers are using algorithms to explore the effects of absorbing aerosols in the stratosphere, especially sulfate aerosols associated with large volcanic eruptions.

Based on these processes, a minimal band set for atmospheric correction can be defined. First, bands should be positioned to avoid specific absorption features in the atmosphere such as water vapor and oxygen. Second, at least two bands with some minimum spectral separation are necessary to characterize the spectral trends with sufficient accuracy. Lastly, bands should be placed in the near infrared (NIR) as noted above. A recent report by the International Ocean Color Coordinating Group (IOCCG, 1998) proposes that bands near 865 and 750 nm, with perhaps a third band near 710 nm, would be suitable.

In-Water Optical Properties

Differentiation of all possible dissolved and particulate materials in the ocean would require far more bands with high spectral resolution in the visible wavelengths than are realistically possible. Therefore this discussion focuses on the primary properties of interest: phytoplankton chlorophyll, colored dissolved organic matter (CDOM), and suspended sediment. Sathyendranath et al. (1994) examined these issues for waters with high levels of suspended and dissolved material that are not correlated with phytoplankton abundance, as are often found in coastal zones and occasionally in the open ocean (Nelson et al., 1998).

As noted earlier, the shift in ocean color as chlorophyll concentration increases is most pronounced around 445 nm, compared with the weak absorption around 560 nm. Atmospherically corrected water-leaving radiances from these two bands are combined in a ratio. This form is used because the sources of variability and error in the radiances, such as changes in the scattering and backscattering coefficients or the impact of bidirectional reflectance, are greatly reduced when band ratios are used. Although the basic band pair of 443 and 560 (with bandwidths of 20 nm) meets most of the requirements, the addition of a third band at 490 nm (which is easier than the 443-nm band to correct for atmospheric effects) will greatly improve the overall performance, especially for quantifying the effects of suspended sediments.

A band that can distinguish between the optical signatures of chlorophyll and CDOM is also required. The divergence between these signatures occurs at wavelengths of less than 440 nm. Because atmospheric correction in the near-ultraviolet is extremely difficult, a compromise band has been used near 410 nm. Again, band ratios are used to estimate CDOM concentrations, much as with chlorophyll.

Other Properties

Some types of phytoplankton have very distinct spectral characteristics that may allow them to be identified even with limited spectral coverage sensors such as SeaWiFS and MODIS. *Trichodesmium*, a cyanobacterium that can fix atmospheric nitrogen, is usually found near the ocean surface and is characterized by high backscatter at 550 nm. Its unique pigment composition can be identified in CZCS imagery (Subramanian and Carpenter, 1994), and new algorithms are being developed for SeaWiFS that will quantify their abundance. *Nodularia*, another cyanobacterium, can also be distinguished in remote sensing of the ocean using a similar approach (Kahru et al., 1994). Coccolithophores, which produce bright calcium carbonate platelets, also change the spectral composition of water-leaving radiance (Balch et al., 1991; Brown and Yoder, 1994). Because they often occur in immense blooms, coccolithophores are easily recognized in ocean color imagery. New algorithms for SeaWiFS and MODIS have been developed to estimate the abundance of coccolithophore platelets.

Beyond estimates of standing stocks, ocean color data are being used to formulate estimates of some important ocean fluxes. The most important of these is primary productivity. A comprehensive review of these models can be found in Behrenfeld and Falkowski (1997). The basic models use estimates of incoming solar radiation, chlorophyll, and ocean temperature to estimate photosynthetic rates. The primary difference between these models is in the way the light response is implemented; one class resolves the vertical structure of the fields, and the other uses vertically integrated fields. The best of these models explain roughly 50 percent of the variability observed in field measurements, and model estimates are within about a factor of two of the actual measurements. However, the quality and the spatial distribution of in situ measurements have their own sources of error, so the satellite-based estimates may actually be somewhat better than would be indicated by these studies.

Productivity models rely on simple estimates of quantum yield, which quantifies the conversion of absorbed light energy into chemical energy. Light absorption can be relatively well described using estimates of chlorophyll, and most of the variability in the productivity models arises from variations in quantum yield. Changes in species composition and nutrient availability are largely responsible for the variability in quantum yield. Measurements of chlorophyll fluorescence yield from satellites may improve productivity estimates because of the generally close relationship between fluorescence yield and quantum yield. Next-generation ocean color sensors, such as MODIS, Global Imager (GLI), and Medium-resolution Imaging Spectrometer (MERIS), include the necessary fluorescence bands at 670, 683, and either 700 or 750 nm with high SNR (Letelier and Abbott, 1996).

Variations in water transparency are critical to predicting the depth of the upper ocean mixed layer, as they determine how solar energy is trapped as a function of depth (Denman and Miyake, 1973). Given that the mixed layer plays a critical role in the flux of energy between the atmosphere and the deep ocean, estimates of mixed-layer depth are important for quantifying Earth's energy budget.

Operational Issues

The spectral features of interest in the ocean are narrow, which leads to a narrow bandwidth requirement for ocean color measurements. Typically, 20 nm is sufficient in the visible range, although the fluorescence bands need to be slightly narrower to avoid atmospheric absorption features. The NIR bands for atmospheric correction can be wider, for example in the 30 to 40 nm range. With high SNR requirements for ocean color, it is also essential that band position be well known and stable over the life of the mission.

Calculations of sensor performance must be based on typical values of water-leaving radiance. In the past, these calculations were made at values for top-of-the-atmosphere (TOA) radiances, which in the case of ocean color are up to ten times higher than the signal of interest. Obviously this can lead to spuriously high SNR values. Table 5.1 shows values for noise-equivalent delta radiance (NEΔL) and bandwidths full width at half maximum (FWHM) for SeaWiFS. These NEΔL values are based on typical water-leaving radiances.

The atmospheric correction and in-water algorithms described above assume that there are no other sources of radiance reaching the sensor. In the case of ocean measurements, two potential sources must be discussed: sun glint and ocean whitecaps. Sun glint is the specular reflection of sunlight off the ocean surface. The size and exact placement of sun glint in a satellite image depend on (1) wind speed (through its forcing of capillary waves) and (2) satellite viewing geometry. CZCS and SeaWiFS are tilting sensors that can look 20 degrees fore or aft to avoid the glint patch. MODIS cannot tilt, and the glint patch must be masked out. Breaking waves create foam and whitecaps, obviously extremely bright targets on a dark ocean. They are not perfectly white reflectors (Frouin et al., 1996). Early models of breaking waves suggested that they would be an important component of the satellite-sensed radiance at wind speeds greater than 10 m s^{-1}. Recent studies suggest that the effect is much smaller and may not become significant until wind speeds exceed 25 m s^{-1}.

As sensor performance improves, processes that once were in the noise of a particular sensor may be part of the signal in its successor. One might question which science requirements are driving the increase in SNR. The basic reason is the low signal emerging from the ocean. For measurements of phytoplankton biomass, we are only interested in those photons that penetrate the sea surface and are backscattered after interaction with suspended particulates (phytoplankton). Thus high SNR will increase our ability to discriminate water types, especially those at low chlorophyll concentration, which make up nearly half of the world's oceans.

TABLE 5.1 SeaWiFS Performance Specifications for Eight Channels, Including Bandwidth Full Width at Half Maximum (FWHM) and the Noise-Equivalent Delta Radiance (NEΔL)

Wavelength	Bandwidth (nm)	NEΔL
412	20	9.2
443	20	7.7
490	20	5.6
510	20	4.9
555	20	4.3
670	20	3.1
765	40	1.9
865	40	1.5

Future Directions

The basic measurements of phytoplankton chlorophyll, CDOM, and suspended sediment fulfill many of the fundamental requirements for global ocean biological studies. Missions such as SeaWiFS and MODIS fulfill these requirements at the necessary temporal and spatial scales. However, there are emerging scientific issues that will result in new requirements. Phytoplankton optical properties vary over the diel cycle, as does cloud coverage. Present and planned satellite missions are clustered around the time period between 0930 and 1200 (except for MODIS on PM-1, which will sample areas obscured by sun glint in the AM-1 MODIS imagery). Orbit crossing times could be spread more widely around local solar noon (e.g., 1000, 1200, and 1400) to investigate diel variability. This would be especially important for measurements of chlorophyll fluorescence (used to estimate quantum yield), which has a significant diel signal. As noted above, fluorescence represents a new measurement type, but the possibilities for improved estimates of primary productivity using fluorescence-based quantum yield look promising.

Hyperspectral remote sensing has been used to study nearshore environments, partly because of the complexity of the optical signal. Early studies such as that of Campbell and Esaias (1983), while not hyperspectral, exploited spectroradiometer data to derive chlorophyll retrievals based on spectral curvature. More sophisticated algorithms (Lee et al., 1994) have been applied to Airborne Visible and Infrared Imaging Spectrometer (AVIRIS) data to derive chlorophyll and CDOM concentrations, as well as bottom depths in shallow waters. As discussed above, the existing algorithms for the interpretation of remotely sensed ocean color signals are based on empirical correlations between the ratios of water-leaving radiances at only a few different wavelengths. These algorithms use a simplified parameterization of the composition of seawater in terms of chlorophyll concentration alone.

A more rigorous approach to understanding remote sensing reflectance begins with determining the inherent optical properties of each of the various optically significant components of seawater. That knowledge enables one to model and understand the roles played by each component in determining the bulk inherent and apparent optical properties of the ocean, including remote-sensing reflectance. Such an approach leads naturally to the development of mechanistic remote-sensing models rather than to correlation models derived from statistical analysis of (usually incomplete) field data. These hyperspectral algorithms may allow us to characterize the types of flora in the ocean based on their pigment composition.

The study of coastal ocean processes will require far more intensive spatial and temporal sampling than the open ocean because of its small characteristic scales. For example, tidal forcing is an important component of the coastal environment, and Sun-synchronous orbits will shift this high-frequency variability into lower frequencies. Geostationary platforms that could sample small regions every 15 minutes could be used to resolve such key processes.

OBSERVING STRATEGY

Time and Space Sampling Requirements

The time and space variability of the ocean must be convolved with the biological time scales to derive an appropriate sampling strategy. Using ship, buoy, and drifter observations, the characteristic time scales for phytoplankton are on the order of 3 to 4 days (with shorter scales in more productive areas, such as the coastal ocean). When combined with typical circulation properties, this leads to characteristic spatial scales of about 20 to 30 km. Given the patterns of clouds that will obscure a large fraction of any one image (average cloudiness over the ocean is about 70 percent), we must have higher resolution to obtain this effective resolution. Chelton and Schlax (1991) have shown that for typical patterns of cloudiness, one can expect to achieve temporal resolution with reasonable errors of about 3 weeks.

Although this is longer than the characteristic scale noted earlier, this does not mean that useful information cannot be obtained from typical satellite sampling of 1 km (nadir resolution) with 2-day global repeat coverage. Instead, the effective resolution requirement depends on the scientific questions being addressed. The Chelton and Schlax (1991) analysis assumes that spatially and temporally uniform error fields are required. This may be the case for science questions involving long time series of global processes, but many mesoscale studies, such as the effect of coastal upwelling, may not have such stringent requirements. Thus the nominal 1-km, 2-day global coverage sampling strategy is a reasonable compromise for a wide variety of scientific problems.

The last part of any sampling strategy is the resilience to temporal gaps in the data record. That is, must continuity be maintained absolutely? Aside from operational requirements, which are not the focus here, most science questions involving ocean color can tolerate some gaps in the record. Given that the dominant ocean climate signal is the ENSO, which recurs approximately every 3 to 5 years, a gap longer than about 6 months runs a serious risk of missing this critical phenomenon.

NASA's Plans

Since the success of CZCS (launched in 1978), NASA has pursued many avenues to obtain science-quality ocean color measurements on a global basis. In 1997, Orbital Sciences, Inc., successfully launched SeaWiFS aboard Orbimage-2 as part of a data purchase agreement with NASA. SeaWiFS should provide measurements through 2002. With its bilinear gain, SeaWiFS will collect data over both the dark ocean and the bright land. MODIS, launched in 1999 aboard the Earth Observing System (EOS) AM-1 platform, is a multipurpose sensor, measuring radiance in the visible portion of the spectrum for both land and ocean applications as well as infrared measurements for surface temperature, atmospheric sounding, and cloud imaging. MODIS has considerably higher SNR than SeaWiFS and also has some 250-m resolution bands for land-surface imaging. Because MODIS does not tilt, a second MODIS will be launched in 2000 on PM-1 to provide global, 2-day coverage, albeit at different orbit crossing times. Given the high level of science involvement in both the SeaWiFS and MODIS missions, these sensors meet the basic science requirements and provide new capabilities to explore new scientific questions.

In 1997, NASA began to reformulate its plans for the second series of EOS missions. Currently, NASA does not have any concrete plans for ocean color measurements after MODIS on PM-1, although it is likely that some plan will emerge during the second series planning. The question remains whether this mission will meet the science requirements. Moreover, there are new directions that the ocean science community wishes to pursue, especially as the U.S. Global Change Research Program (USGCRP) begins to focus on linkages between ecosystem processes and climate change. With the success of SeaWiFS and the anticipated success of MODIS, there will be considerable pressure to maintain at least MODIS-quality ocean color measurements.

NPOESS Plans

The Department of Defense and NOAA have developed a set of operational requirements for ocean color measurements as part of the environmental data records (EDRs) process (see Table 5.2). These EDR requirements will be met by the Visible and Infrared Imaging Sensor (VIIRS); it has not been decided whether VIIRS will consist of a single package like MODIS, or if it will consist of separate sensor modules. These requirements were reviewed briefly in a climate workshop (NOAA, 1997).

The present threshold and objective requirements for ocean color are difficult to evaluate for several reasons. First, only one of the planned National Polar-orbiting Operational Environmental Satellite System (NPOESS) orbits will be able to make measurements of sufficient quality (crossing time 1330); 0530 and 0930 are simply too early in the day. Second, the requirements are reported in terms of chlorophyll rather than as radiances. Since chlorophyll errors will be a function of the various atmospheric correction and in-water algorithms, it will simplify sensor comparisons if they are based on normalized water-leaving radiances (which have the atmospheric correction) and TOA radiances (which do not). Third, many important sensor specifications are absent: for example, spectral bandwidth, band stability, digitization rate, bright target recovery, band-to-band registration, and so forth are not specified. This was done largely because the NPOESS EDR process was based on final data products, not on a specific sensor design. While this approach may result in a lower-cost sensor and also does not lock NPOESS into a particular technical implementation, it is based on an assumption that the quality of a data product is independent of the sensor design. As was learned with SeaWiFS, this may not be a realistic assumption. Fourth,

TABLE 5.2 NPOESS Environmental Data Records for Ocean Color

Ocean Color/Chlorophyll	Threshold	Objective
Horizontal resolution		
Global, worst case	2.6 km	1 km
Regional, worst case	1.3 km	0.1 km
Mapping accuracy		
Global, worst case	3 km	0.5 km
Regional, worst case	3 km	0.1 km
Measurement range	0.05-50 mg/m^3	0-100 mg/m^3
Measurement precision	20%	10%
Measurement accuracy	±30%	±30%
Refresh	48 h	12 h
Turbidity		
Sensing depth	Surface	
Horizontal resolution	1.3 km	0.25 km
Mapping accuracy	TBD[a]	0.5 km
Measurement range	TBD	0-100 mg/l
Measurement precision	TBD	0.1 mg/l
Measurement accuracy	±30%	±0.1 mg/l
Refresh	48 h	24 h

[a]TBD, to be determined.
SOURCE: IPO NPOESS (1996). The updated IORD and other documentation related to the NPOESS program are available online at <http://npoesslib.ipo.noaa.gov/ElectLib.htm>.

there are no explicit requirements for long-term stability. Although there is an implicit expectation that the data product must continue to meet its threshold performance, experience has shown that new platforms are replaced only when critical sensors fail. The slow degradation of a noncritical data product such as ocean color may result in a long period of scientifically useless data. More importantly, there is no explicit plan for periodic calibration and validation, as there is with sensors such as SeaWiFS and MODIS. The committee's assessment of the current status and future NPOESS plans for the observational measurement of ocean color can be found in Box 5.1.

International Plans

Other nations are planning ocean color sensors; notable among these are the Global Imager on the Japanese ADEOS-2 spacecraft and MERIS on the European ENVISAT. Both of these missions are strongly science-driven, and although there are differences in performance and objectives, both are similar to MODIS. Figure 5.1 compares the performance of SeaWiFS, the Japanese Ocean Color Temperature Scanner (OCTS) on ADEOS-1, MODIS, GLI, and MERIS.

Although GLI and MERIS are highly capable sensors, it will be difficult to use them as a basis for developing a strategy for long time series of ocean color for studies of climate-related processes, given the uncertainty of international planning. Therefore, the most prudent strategy is one based on a combination of research and operational missions.

DATA PRODUCTS

The critical data products are discussed earlier in this chapter. The fundamental set includes phytoplankton chlorophyll, CDOM, and suspended sediment. To obtain these products requires a complicated set of atmospheric correction algorithms. Because these two classes of algorithms are still the subject of research, it is difficult to

Box 5.1
Findings

1. Some of the basic science requirements for ocean color remote sensing will be met by the VIIRS sensor on NPOESS. The EDRs for chlorophyll and suspended sediment are congruent with those developed for SeaWiFS. However, there is no EDR for CDOM (see Table 5.1), an important component of the ocean optical environment and its biogeochemistry. Moreover, there are many other aspects of overall system performance that must be met.

2. The plans for VIIRS are difficult to evaluate because many critical elements are incomplete or not yet determined. This stems from the EDR approach, which proceeds from the definition of the data product requirements rather than specifications of the technology. These missing elements include calibration and validation, data archiving, sensor characterization, algorithm development, and technology insertion. Climate research requires more than hardware in orbit; it requires a complete, integrated program of sensor engineering and testing, as well as scientific research.

3. The first series of EOS sensors will explore new capabilities in ocean color remote sensing. The first series of EOS platforms (AM-1 and PM-1) will meet the basic science requirements for ocean color, as will SeaWiFS. MODIS, with additional bands and better performance, will support the development of new data products and testing of new concepts. For example, chlorophyll fluorescence bands, which are part of every upcoming ocean color sensor (MODIS, MERIS, and GLI), will not be included in VIIRS. These bands will significantly improve estimates of primary productivity in the upper ocean. Although NASA may continue such measurements in a research context, there is no mechanism to insert these new requirements into the operational NPOESS system.

4. The present plans (as they have been defined) do not ensure that a continuous time series of ocean color data will be collected that is suitable for climate research. There is the possibility of a gap between the first series of EOS missions and the first NPOESS mission, and it is probable that any "gap filler" will be less capable than MODIS. However, if AM-1 lasts only until 2005, then there may be a significant drop in coverage until the launch of VIIRS in 2009. This gap may be eliminated by a sensor on the second EOS series, but the quality of such a sensor and its associated data products is still being defined. NASA has committed to continuity of ocean color measurements, perhaps through an early flight of VIIRS, which is being studied as part of a joint NASA/NOAA bridging mission to fly in the 2005 to 2009 time frame.

5. While the ocean color measurements in NPOESS may be useful for long time series, considerably more planning and coordination are required. If dynamic continuity is to be achieved, programs such as Sensor Intercomparison and Merger for Biological and Interdisciplinary Oceanic Studies (SIMBIOS) must be extended and maintained to include the operational missions as well as the research missions. Required activities include adequate prelaunch characterization and testing, as well as ongoing data product validation and analysis. Improvement is needed in coordinating NASA's technology innovation and climate research with the Integrated Program Office's (IPO's) commitment to long time series of operational missions. NASA's technology plans are moving so rapidly that it is unlikely that any new technology will become part of NPOESS, given the mismatch in schedules. An explicit plan must be developed for ocean color research to transfer to the operational systems.

6. The fact that ocean ecosystems change on decadal scales in response to climate change makes their observation especially challenging. The commitment of NASA and IPO/NPOESS to long time series is a critical component of an observing strategy, but the present plans are incomplete. It is not simply a case of defining better requirements for NPOESS ocean color observations, though this is needed. A complete program of research, analysis, and technical innovation, with a long-term commitment to the measurements, is required. There is an opportunity to achieve these scientific goals without an enormous financial investment. The research missions begun by NASA have established a strong scientific basis for ocean color as well as a sound technical basis. NPOESS represents an opportunity to ensure long-term observations. The challenge will be to blend active research, technology innovation, and continuous measurements to develop an observing system for climate research.

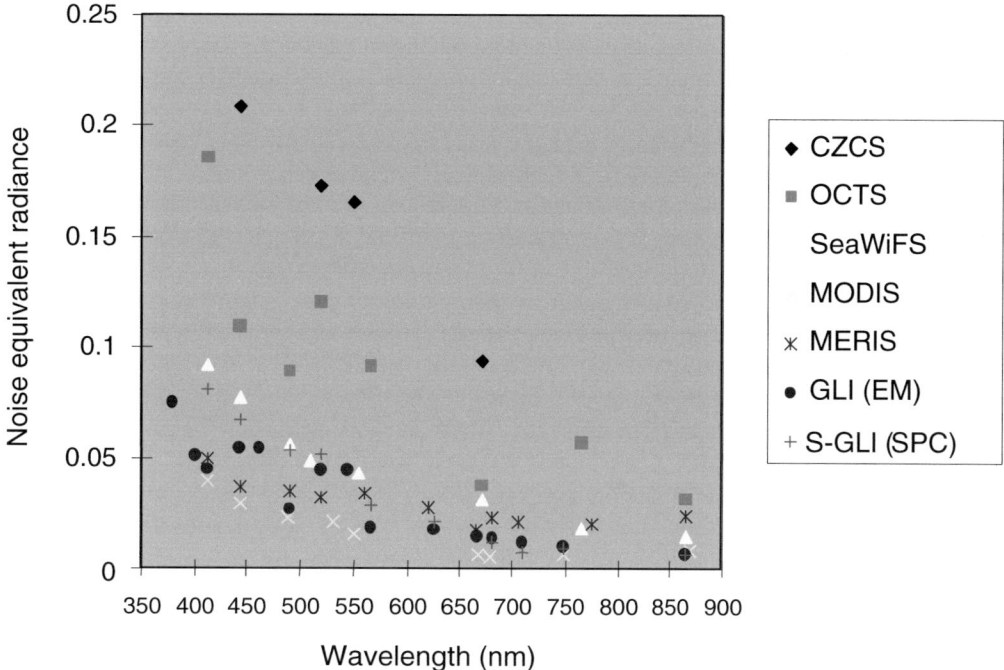

FIGURE 5.1 Noise-equivalent radiance as a function of wavelength for several ocean color sensors. The distribution shows the improvement in sensor performance from the first ocean color sensor (CZCS) to the upcoming sensors of MODIS, MERIS, and GLI. The S-GLI sensor is planned for ADEOS-3. Acronyms are defined in Appendix B.

conceive how they can be "purchased" as part of a sensor package. Ocean algorithms will continue to evolve, especially as the time series lengthen, new problems arise, and scientific understanding improves. Examples of possible improvements are in accounting for absorbing aerosols in the stratosphere, bidirectional reflectance off the ocean surface, and in-water algorithms based on inherent optical properties. Therefore a program of active research to improve and evaluate algorithms must also be continued. NASA has a history of supporting research to improve data sets, although it generally focuses on data sets collected by NASA missions. The NASA/NOAA Pathfinder data sets were a notable exception, with the two agencies collaborating to produce high-quality data sets for Earth science.

One particular area where ocean color algorithms need attention is the estimation of water column productivity. Although fluorescence should improve these estimates, there is no plan within NPOESS to continue these measurements. Therefore, it is probable that productivity estimates will have to rely on indirect estimates of the photoadaptive state. This may include estimates of mixed-layer depth, nutrient concentration through the use of nutrient and temperature climatologies, and so forth. These models may be more robust than simple statistical correlations, but the additional uncertainties that arise from these new parameterizations will pose new challenges.

CALIBRATION AND VALIDATION

SeaWiFS and MODIS have extensive onboard calibration systems to monitor sensor performance. These systems include monitoring of sensor response through lunar and solar calibration, as well as monitoring of spectral stability (for MODIS). Although these systems (and in the case of lunar calibration, orbital maneuvers) will monitor sensor performance, subtle changes may result in significant changes in derived data products. NPOESS will maintain some level of calibration, but neither the requirements nor the approach have been defined.

As discussed in this and earlier chapters, onboard monitoring is necessary but not sufficient. A rigorous program of vicarious calibration through the use of in situ measurements is also necessary. Both SeaWiFS and MODIS have ongoing programs of field studies in support of vicarious calibration. Extensive cruises and moored buoys are part of this program. These programs are necessarily costly, given the need to ensure that the in-water measurements are themselves well calibrated. Thus there are recurring questions about alternative approaches to such vicarious calibration. However, there is no suitable alternative if the requirement is to maintain data quality for climate research. NPOESS has not defined a plan for vicarious calibration.

Neither NASA nor NPOESS has yet committed to maintaining dynamic continuity between sensors. After AM-1 and PM-1, it is not clear what type of ocean color sensors NASA will fly, or how they will overlap with the first EOS series. The transition to the operational VIIRS sensors on NPOESS is also not yet planned. The risk is that an enormous investment will be made, yet it will not be possible to intercalibrate the data sets. Dynamic continuity was challenging for the Pathfinder data sets, where identical sensors were flown and the sensors and their associated algorithms were simpler. Given the low signal, ocean color requires precise and accurate knowledge of calibration. Ground networks (including process studies, moorings, and drifters) provide well-calibrated data sets that are essential for interrelating successive satellite sensors. Lastly, insertion of new technology developed by NASA or NPOESS must be part of any continuous time series. It is not apparent how new technology infusion will be accommodated in NPOESS after the final design for VIIRS is selected.

NASA has recently started the SIMBIOS (Sensor Intercomparison and Merger for Biological and Interdisciplinary Oceanic Studies) program, which is designed to gather in situ information necessary to develop a consistent time series of ocean color from multiple satellite sensors. Although it initially is focused on U.S. sensors, the IOCCG (working under the auspices of the Committee on Earth Observation Satellites) is studying a possible international program that would incorporate all the ocean color research missions. There is no firm plan to continue this activity beyond 2001.

Programs such as SIMBIOS require extensive prelaunch sensor characterization studies. Tests for SeaWiFS and MODIS included polarization, spectral response, scan mirror reflectivity, and others. These activities are costly, and there is frequently considerable pressure to eliminate or scale back such tests to lower costs or preserve schedule. However, one lesson learned from any time series is that such characterization is essential. Moreover, the data from such tests must be preserved and documented with as much care as the original satellite data. Since such tests are part of sensor construction, they are sometimes treated simply as engineering data and are preserved only in written form. NPOESS contractors are supposed to maintain this information in an accessible form. For MODIS, these tests are conducted with the cognizance of the science team.

Validation consists of estimating the quality of the derived data products, including the effects of sensor noise and processing algorithms. Estimation of the impact of sampling errors is also part of the validation process. SIMBIOS is supporting some activities in this regard, although the focus is on the algorithm part of the error budget. One of the MODIS team activities is assessment of algorithm errors as well. There are no identified plans for data product validation as part of NPOESS.

EVOLUTION STRATEGY

NASA has added three programs to develop and test new technology. Although there are no specific plans yet in the area of ocean color, some proposed candidate technologies include measurements from geostationary orbit to study high-frequency events, hyperspectral measurements to examine phytoplankton pigment groups, fluorescence imagers to improve productivity models, and lidar to study mixed layers and pigment fluorescence. The Integrated Program Office for NPOESS (IPO/NPOESS) is planning to reserve some platform resources to accommodate new technology, but this plan is not well defined at present, nor is there a mechanism to exploit these resources.

NASA has moved strongly toward a program of rapid technological development in order to lower costs and develop more capable sensors in response to improvements in scientific understanding. Many of these missions have short time scales, completing the planning and design phase in 1 to 2 years, followed by a relatively short mission. NASA technology programs such as the New Millennium Program and the Instrument Incubator Program are not focused on improving the capabilities of NPOESS. If NPOESS planning takes 5 or 6 years, it is not clear how NASA technology will be incorporated into NPOESS. Operational missions require several years of proven spaceflight to increase confidence in the sensor design as well as demonstrate the utility of the data set.

REFERENCES

Balch, W.M., P.M. Holligan, S.G. Ackleson, and K.J. Voss. 1991. Biological and optical properties of mesoscale coccolithophore blooms in the Gulf of Maine. Limnol. Oceanogr. 36: 629-643.

Behrenfeld, M.J., and P.G. Falkowski. 1997. Photosynthetic rates derived from satellite-based chlorophyll concentration. Limnol. Oceanogr. 42: 1-20.

Brown, C., and J.A. Yoder. 1994. Coccolithophorid blooms in the global ocean. J. Geophys. Res. 99: 7467-7482.

Campbell, J.W., and W.E. Esaias. 1983. Basis for spectral curvature algorithms in remote sensing of chlorophyll. Appl. Opt. 22: 1084-1093.

Chelton, D.B., and M.G. Schlax. 1991. Estimation of time-averages from irregularly spaced observations: With application to coastal zone color scanner estimates of chlorophyll *a* concentrations. J. Geophys. Res. 96: 14669-14692.

Denman, K.L., and M. Miyake. 1973. Upper layer modifications at Ocean Station Papa: Observations and simulation. J. Phys. Oceanogr. 3: 185-196.

Frouin, R., M. Schwindling, and P.-Y. Deschamps. 1996. Spectral reflectance of sea foam in the visible and near-infrared: In situ measurements and implications for remote sensing of ocean color and aerosols. J. Geophys. Res. 101: 14361-14371.

Gordon, H.R., and A.Y. Morel. 1983. Remote Assessment of Ocean Color for Interpretation of Satellite Visible Imagery. A Review. New York: Springer-Verlag.

Integrated Program Office (IPO), National Polar-orbiting Operational Environmental Satellite System (NPOESS). 1996. Integrated Operational Requirements Document (IORD) I. Joint Agency Requirements Group Administrators. 61 pp. + figures.

International Ocean Color Coordinating Group (IOCCG). 1998. Minimum requirements for an operational ocean-colour sensor for the open ocean. IOCCG Rept. No. 1, Dartmouth, Nova Scotia, Canada.

Kahru, M., U. Horstmann, and O. Rud. 1994. Satellite detection of increased cyanobacteria blooms in the Baltic Sea: Natural fluctuation or ecosystem change. Ambio 23: 469-472.

Kirk, J.T.O. 1994. Light and Photosynthesis in Aquatic Ecosystems. New York: Cambridge University Press.

Lee, Z., K.L. Carder, S.K. Hawes, R.G. Steward, T.G. Peacock, and C.O. Davis. 1994. Model for the interpretation of hyperspectral remote-sensing reflectance. Appl. Opt. 33: 5721-5732.

Letelier, R.M., and M.R. Abbott. 1996. An analysis of chlorophyll fluorescence algorithms for the Moderate Resolution Imaging Spectrometer (MODIS). Remote Sensing Environ. 58: 215-223.

National Oceanic and Atmospheric Administration (NOAA). 1997. Climate Measurement Requirements for the National Polar-orbiting Operational Environmental Satellite System (NPOESS), Workshop Report, Herbert Jacobowitz (ed.), Office of Research Applications, NESDIS-NOAA, February. 77 pp.

Nelson, N.B., D.A. Siegel, and A.F. Michaels. 1998. Seasonal dynamics of colored dissolved material in the Sargasso Sea. Deep-Sea Res. 40: 931-957.

Sathyendranath, S., F.E. Hoge, T. Platt, and R.N. Swift. 1994. Detection of phytoplankton pigments from ocean color: Improved algorithms. Appl. Opt. 33: 1081-1089.

Subramanian, A., and E.J. Carpenter. 1994. An empirically-derived protocol for the detection of blooms of the marine cyanobacteria *Trichodesmium* using CZCS imagery. Int. J. Remote Sensing 15: 1559-1569.

6

Soil Moisture

INTRODUCTION

The existence, quantity, and nature of all life forms on our planet are closely linked to the distribution and phase of water in Earth's biosphere. Soil moisture is a fundamental parameter of the terrestrial environment; its spatial distribution and temporal variation are crucial ingredients of hydrologic, ecologic, and climatic models, on regional and global scales. It is estimated that only 4 percent of Earth's water is stored terrestrially. Of this, approximately 25 percent is subsurface water and the remainder is stored primarily in the solid phase as snow and ice. Though of relatively small total quantity, water in the terrestrial regime plays an active role in the global hydrologic cycle and is estimated to contribute 21 percent of global precipitation and 14 percent of global evapotranspiration.

The horizontal and vertical distributions of soil moisture exert control over the partitioning of incoming radiant energy into latent and sensible heat fluxes via evaporation and transpiration by plants. Soil moisture also affects soil albedo. Near-surface soil-moisture distribution helps to determine the fate of precipitation as surface runoff or as percolation down into the soil column. The timing and magnitudes of these horizontal and vertical transports of liquid water are important determinants in the geochemical cycling of nutrients, the availability of soil water for plant growth, the recharge of aquifers and reservoirs, and soil engineering properties. Excessive percolation can lead to leaching of soil nutrients, as is common in the humid tropics. Excessive runoff can lead to catastrophic flooding. The recent flooding and landslides in Central America caused by hurricane Mitch may have been exacerbated by antecedent drought conditions related to El Niño/Southern Oscillation (ENSO).

Soil-moisture information is also useful for commercial applications associated with farming practices and irrigation. The annual growth rate of world agricultural production has slowed from about 3 percent in the 1960s to less than 2 percent (FAO, 1995). This trend is expected to continue as more marginal lands are brought into production. Productivity is generally limited by either nutrient or moisture availability. Much of the focus of the green revolution was on optimizing nutrient availability, with increased use of irrigation to redress moisture deficiencies. Currently 19 countries are listed as "water scarce," with less than 100 cubic meters of fresh water available per capita, and the number is expected to increase to more than 46 countries over the next 30 years (NOAA, 1997). This trend, based on population growth, will be compounded by increasing agricultural demands for water due to continued agricultural expansion into semiarid and arid regions dependent on irrigation. The status and availability of soil moisture are expected to be of increasing geopolitical significance in such regions as the Middle East and sub-Saharan Africa.

Despite its unquestioned significance and recognition by science working groups that it is among the most important environmental parameters that NASA's suite of sensing satellites should monitor on a global scale, no means are available today to map soil moisture on any scale, nor are there any specific plans to do so in the near future. Soil moisture has not been widely applied as a variable in any land process models for two primary reasons. It is a difficult variable to measure on a consistent and spatially comprehensive basis. It also exhibits great spatial and temporal variability; hence point measurements have little meaning. Consequently, soil moisture has not been used as a measurable variable in current hydrologic, climatic, agricultural, or biogeochemical models due to a lack of appropriate data. Although soil moisture is listed by NOAA as a critical climate environmental data record (EDR) (NOAA, 1997), it is the only EDR for which no threshold and objective values are provided. The committee's assessment of the current status and future NPOESS plans for observing and measuring soil moisture from space is discussed in Box 6.1.

BASIC SCIENCE ISSUES

Because of its ubiquity, there are numerous potential science applications for frequent and spatially comprehensive measurements of soil moisture. Most of these fit under the following four science issues:

• Understanding the role of surface soil in the partitioning of incoming radiant energy into latent and sensible heat fluxes at a variety of scales, from the mesoscale to general circulation model (GCM) scale;
• Understanding the relationship between the moisture in the top 5 cm of soil that is observable by microwave techniques and the total profile (1 m or more) of soil moisture that is accessible to plants and is available for transpiration to the atmosphere;
• Understanding how spatial and temporal patterns of soil moisture are related to the physical and hydrological properties of soils; and
• Understanding how the spatial and temporal patterns of soil moisture can be used to improve our ability to model runoff at a variety of scales and adapt hydrologic models to areas of differing climate, biomes and soils, and geology.

With a potential for measuring soil moisture demonstrated, how might society use such soil moisture measurements? As in the science issues, there are four general areas in which routine measurements of soil moisture could have major impacts on day-to-day life:

• Improving medium-range weather forecasting by incorporating measured soil moisture on a 30-km grid daily;
• Improving on-farm irrigation scheduling and efficiency, and improving crop yield modeling for domestic and foreign areas, among other agricultural applications, at scales of 10 m to 100 m and 1 to 3 days.
• Better quantifying water use, storage, and runoff to monitor existing resources and to assist decision makers in allocating limited resources or coordinating relief efforts in times of flooding, at scales of 100 m to 1 km and daily or on demand.
• Improving climate models, particularly for annual and interannual variability, so that they represent the land surface hydrologic processes accurately. Measured soil moisture can be used as a state variable and as a validity measure for GCMs. The scales of interest are 1 to 10 km, possibly averaged to coarser resolution, at time scales of 1 to 7 days.

Status of Soil Moisture Sensing

The status of soil-moisture sensing may change somewhat after the European Space Agency (ESA) launches ENVISAT in June 2001 and the National Space Development Agency (NASDA) of Japan launches PALSAR in

> **Box 6.1**
> **Findings**
>
> Soil moisture would be a valuable input to the current hydrologic, climatic, agricultural, and biogeochemical models if it could be reliably measured at the appropriate spatial scales and over time intervals relevant to the processes of interest. A brief summary of the importance of soil moisture is given below, modified from Evans et al. (1995). An important role of the land surface component in general circulation models (GCMs) is to partition incoming radiative energy into sensible and latent heat fluxes. Soil moisture is known to be the major determining factor. A number of modeling studies have demonstrated the sensitivity of soil moisture anomalies to climate (Walker and Rowntree, 1977; Rind, 1982; Shukla and Mintz, 1982; Delworth and Manabe, 1988). Soil moisture is found to be an important forcing function, second only to sea surface temperature at mid-latitudes, and it becomes the most important forcing function in the summer months.
>
> The role of soil moisture is equally important on smaller scales. Recent studies with mesoscale atmospheric models have demonstrated sensitivity of circulation and precipitation to spatial gradients of soil moisture. Fast and McCorcle (1991) have shown that soil moisture gradients can generate thermally induced circulation similar to sea breezes. Chang and Wetzel (1991) concluded that the spatial variations of vegetation and soil moisture affect surface baroclinic structures through differential heating, which in turn indicate the location and intensity of surface dynamic and thermodynamic discontinuities necessary to the development of severe storms. In yet another study, Lanicci et al. (1987) showed that dry soil conditions over northern Mexico and variable soil moisture conditions over the southern Great Plains dynamically interacted to alter prestorm conditions and subsequent convective rainfall patterns.
>
> It is still unclear whether the spatial distribution of soil moisture collected at regional scales is useful for GCM and mesoscale modeling. One positive indicator is the recent study by Betts et al. (1994) during the summer of 1993 in the U.S. Great Plains region, showing that the use of current soil moisture measurements to initialize a forecast model can lead to improved rainfall predictions. Extreme wetness, in comparison with climatological average soil moisture, clearly was a factor in the effect. For more normal conditions, soil moisture anomalies will vary with the spatial and temporal scales of rain events—scales that may be meaningful to four-dimensional data assimilations and mesoscale modeling.
>
> Based on these studies and scales ranging from GCM to the mesoscale, it appears that soil moisture will be an important hydrologic variable for hydrometeorological modeling and validation studies.
>
> While the scientific needs are compelling and there are widespread applied needs that could be met, there is no existing satellite capability for measurement of soil moisture, nor are there any specific plans to develop such a sensor either for scientific research or to meet operational needs. The committee notes that microwave sensors (active and passive) have demonstrated a capability to provide near-surface soil moisture information under certain land-cover conditions and at certain spatial scales. There is a clear need for a study to make specific recommendations for such a capability. The committee urges that the study consider both research use and applied operational use, consider the appropriate range of spatial and temporal scales, consider calibration and validation of the sensor and derived information, and consider issues of integration with other operational and research programs. The committee considers it to be within the scope of NASA's overall mission to move forward with such a program in collaboration with NOAA and USDA.

2002. Each of these satellites is configured to carry dual-polarized synthetic aperture radar (SAR), which may allow the mapping of soil moisture for bare soil surfaces and for surfaces with short vegetation cover. To map soil moisture under a wide range of vegetation-cover conditions (but still limited to grasses and cultural vegetation), it is necessary to use a dual-frequency multipolarized SAR. NASA's recently announced LightSAR, a comparatively small, inexpensive spaceborne SAR, may offer the requisite combination of wavelengths and polarizations for extracting soil moisture information from SAR images, for both bare and vegetation-covered terrain.

Among electromagnetic remote-sensing techniques, only active (radar) and passive (radiometry) microwave sensors have shown strong sensitivity to soil moisture content. Although the active and passive microwave approaches have often been regarded as competing or alternative options for sensing soil moisture, they are complementary in terms of what they offer. A dual-frequency, multipolarized SAR can generate high- to moderate-resolution (3 to 300 m) soil-moisture maps of nonforested terrain, whereas a low-frequency (L-band) microwave radiometer is particularly suited for mapping sparsely vegetated terrain at resolution scales in the range of 10 to 20 km.

The next three subsections discuss passive microwave remote sensing as a potential technique for mapping the distribution of soil moisture, active microwave remote sensing, and the complementarity of the two techniques. The three sections primarily address two questions:

1. How well can passive and active microwave sensors monitor soil moisture? and
2. What sensor configurations are needed to achieve it?

These discussions provide the background necessary for considering what soil-moisture estimation capability we might expect from satellite sensors planned for launch in the next few years.

Passive Microwave Sensing of Soil Moisture

Because microwave emission and backscatter by terrain are strongly influenced by vegetation cover, the committee first considers bare-soil surfaces and then vegetation-covered surfaces. For the purposes of this discussion, a soil surface is defined as bare if it is covered with less than 0.1 kg/m^2 of vegetation. For a green field of grass, this biomass level corresponds to a height of about 10 cm. To maximize penetration through the vegetation cover as well as through the soil's top layer, the band of choice for sensing soil moisture is the L-band wavelength (21 to 23 cm), or longer if technologically feasible. At these wavelengths, a biomass level of 0.1 kg/m^2 exercises negligible effects on the emission and backscatter from the underlying soil surface. Thus, the distinction between bare and vegetation-covered surfaces is made from the perspective of the microwave sensor, which is important because under the stated definition the bare-soil category encompasses a significantly larger fraction of Earth's surface than it would under a definition specifying totally barren.

Bare-Soil Surfaces

The emission from a bare-soil surface that would be observed by a microwave radiometer with its antenna beam pointed at the surface is governed by three main physical parameters: (1) the soil-surface temperature (T_s), (2) the volumetric moisture content of the near-surface layer (m_v), and (3) the roughness of the air-soil interface, usually represented by the root mean square (rms) value of the surface height(s) (Choudhury et al., 1978; Jackson, 1993; Ulaby et al., 1986). Soil type, defined by texture and clay mineralogy, also exerts control over the microwave dielectric properties of moist soil (Hallikainen et al., 1985; Dobson et al., 1985). The effective depth of the emitting soil layer varies between 0.1 λ for wet soils and 1.0 λ for a dry-soil medium, where λ is the wavelength. In practice, the effective depth is considered to be 5 cm. After atmospheric corrections, measured values of the brightness temperature observed by the radiometer are converted into estimates of soil moisture after using infrared observations to estimate an effective surface temperature T_s and making corrections for the effects of roughness and soil type on the basis of auxiliary information about the terrain site under observation. Even with auxiliary information, however, it is difficult to estimate surface roughness, which renders the estimation process more or less heuristic. Fortunately, unless the surface is very rough, the error associated with the assignment of the estimated value of surface roughness leads to an acceptable level of error in terms of the estimated value of soil moisture.

Vegetation-Covered Surfaces

The presence of a vegetation layer above the soil surface alters emission in two ways: (1) it attenuates the energy emitted by the soil and (2) it adds an emission component of its own. Part of this component is emitted by the vegetation directly upward into the air; another part is initially emitted downward toward the soil surface and then, upon reflection by the surface, propagates upward through the vegetation layer into the air (Jackson and Schmugge, 1991). The vegetation layer introduces a masking effect, reducing the sensitivity of the microwave radiometer to soil moisture. The magnitude of the masking effect is dependent upon the structure and wet biomass of the vegetation layer. Both the attenuation and emission characteristics of the vegetation layer are wavelength-dependent; from a soil-moisture standpoint, it is desirable to choose as long a wavelength as possible. Because of other considerations, especially the fact that the angular resolution capability of a passive microwave instrument is governed by its antenna dimensions (measured in wavelengths), the consensus in the scientific community has been to designate the L-band as the optimum wavelength for passive microwave mapping of soil moisture. From a 600-km altitude, an L-band satellite-borne radiometer with a 15-m-diameter antenna (or an equivalent thinned-array configuration) would have an instantaneous field of view of about 10 km. When made to scan in a conical configuration at an approximately constant incidence angle of 45°, such a system can cover a swath width on the order of 1,000 km.

Passive Microwave Feasibility Demonstration

Over a period of 4 years, a team of researchers from the U.S. Department of Agriculture, NASA/Goddard Space Flight Center (GSFC), and the University of Massachusetts conducted several related investigations aimed at demonstrating two specific capabilities: (1) that it is possible to build and use a thinned-array antenna configuration for radiometric observations, thereby reducing the weight of the radiometer instrument by about a factor of 5, and (2) that it is possible to use an L-band radiometer for mapping soil moisture "over a range of cover conditions and within reasonable error bounds" (Jackson and LeVine, 1996). The first objective led to the development of an aircraft-mounted system called the Electronically Scanned Thinned Array Radiometer, or ESTAR. After its operational capabilities were demonstrated by comparing its performance to that of a traditional push-broom scanning instrument (Jackson et al., 1993; LeVine et al., 1990), ESTAR was used to achieve the second objective over a test site in Oklahoma known as the Washita site.

The results of the study are summarized in Figures 6.1 and 6.2. The ESTAR instrument was flown over the test site on each of 8 days between June 10 and June 18, 1992. Each brightness temperature image was processed through an algorithm that attempts to correct for surface roughness and vegetation cover (estimated from SPOT data) to generate a soil-moisture image. The eight images displayed in Figure 6.1 followed a period of heavy rains over several weeks that ended on June 9, one day prior to the first ESTAR overflight. Thus, the soil was mostly saturated on June 10. Over the next 10 days, no rainfall occurred, allowing the soil to dry down and the ESTAR observations to display the gradual change (Jackson et al., 1995).

Figure 6.2 contains plots of model-calculated values of soil moisture versus brightness temperature, as well as measured values for selected ground surfaces. Based on this study by Jackson and LeVine (1996), the estimated accuracy of the soil moisture content predicted by the ESTAR instrument is 3.5 percent for bare-soil fields and 5.7 percent for vegetation-covered fields.

Regardless of which imaging technique is used, a passive microwave sensor is expected to generate 1,000-km-wide maps of soil moisture at a resolution of about 10 km × 10 km and at an accuracy of 3 to 6 percent, over the soil moisture range between 5 and 30 percent. With such a swath width, revisit time would be on the order of 3 days. The accuracy, swath width, resolution, and revisit interval of the soil moisture information provided by such a sensor is compatible with data inputs needed for the implementation of hydrologic models of large basins in most environments.

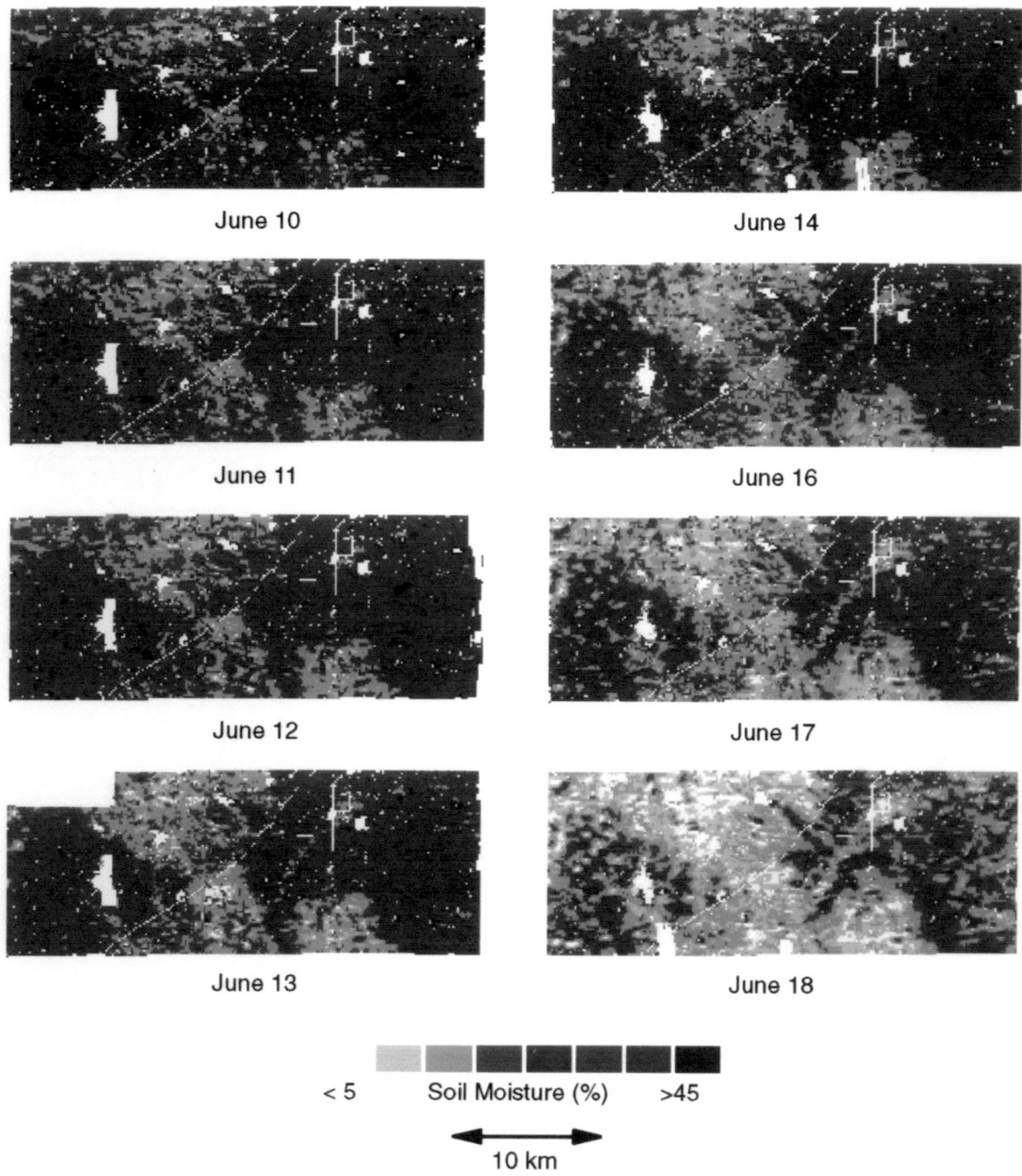

FIGURE 6.1 Surface soil moisture images for Washita '92 (Jackson and LeVine, 1996).

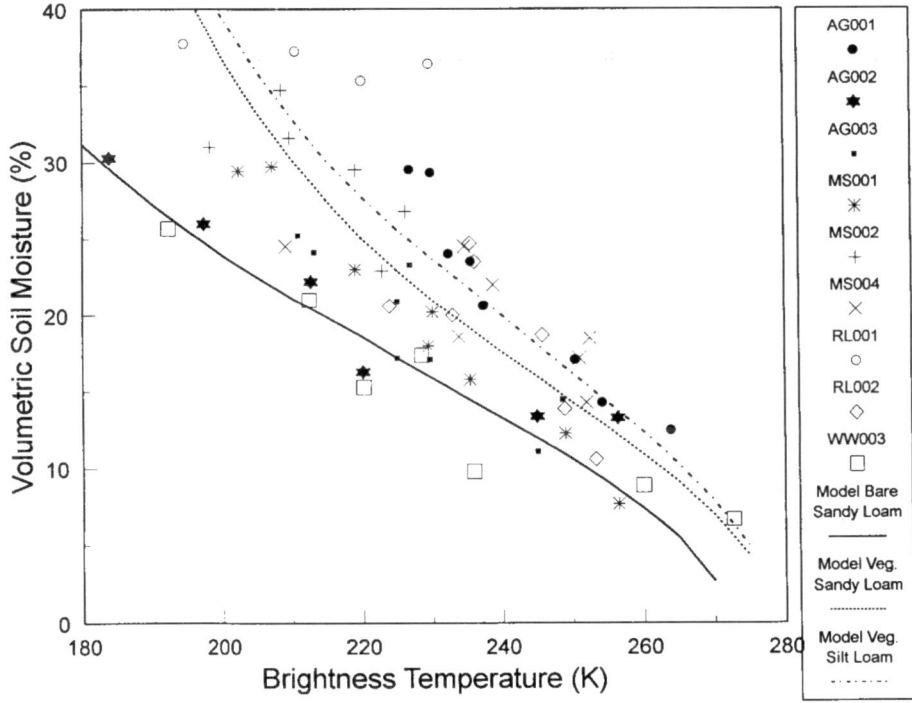

FIGURE 6.2 Ground verification of soil moisture estimation algorithm (Jackson and Le Vine, 1996).

Active Microwave Sensing

Bare-Soil Surfaces

The backscattering coefficient σ° measured by a radar system whose antenna beam is pointed at a bare-soil surface is governed by two physical parameters: the volumetric moisture content m_v and the rms surface roughness s, and to a much lesser extent by soil type.

The functional dependence of σ° on moisture, as well as on roughness, is polarization-sensitive, meaning that the response of σ° to moisture and roughness varies for different combinations of transmit and receive antenna configurations. For a typical observational scenario (incidence angle and polarization), the dynamic range over which σ° varies in response to m_v, varying from dry conditions to high wetness conditions, is comparable to the dynamic range over which σ° varies in response to surface roughness from smooth to very rough (for natural surfaces at L-band). Because σ° is equally sensitive to the soil moisture content m_v and the surface roughness s, attempts by many researchers to use single-channel ERS-1/2 or JERS-1 SAR observations for estimating soil moisture have not met with much success. Conversely, observations by the Jet Propulsion Laboratory (JPL) airborne SAR (AirSAR) and by Shuttle Imaging Radar-C (SIR-C) have both shown that soil moisture can be retrieved with an accuracy of about 3.5 percent when two or more L-band polarization channels are used (Figure 6.3).

Implementation of a SAR soil-moisture inversion algorithm (Dubois et al., 1995; Oh et al., 1992; Ulaby et al., 1996) to SIR-C images of an area in Oklahoma led to the soil-moisture distribution maps shown in Figure 6.4. These figures correspond to relatively wet conditions (20 percent average) for the April 12, 1994, map and to relatively dry conditions (10 percent average) for the April 15 map.

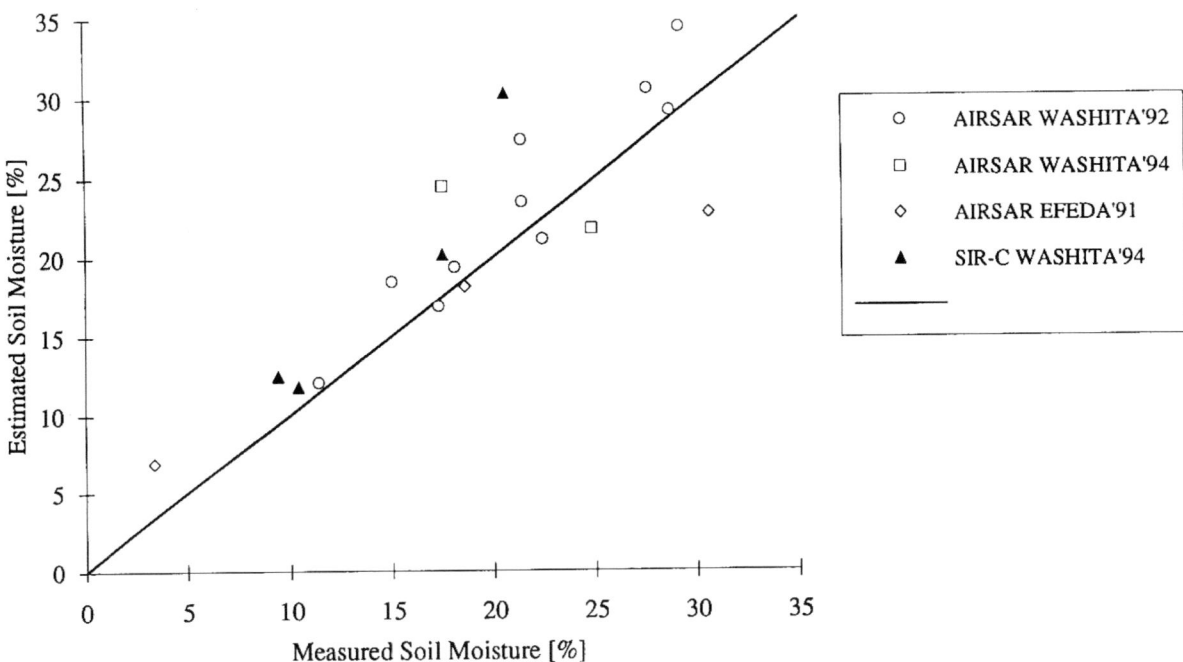

FIGURE 6.3 Comparison of radar-estimated soil-moisture values for bare-soil fields with in situ measurements. The synthetic aperture radar (SAR) observations were extracted from several AirSAR campaigns and from Shuttle Imaging Radar-C overpasses in April 1994 (Ulaby et al., 1996).

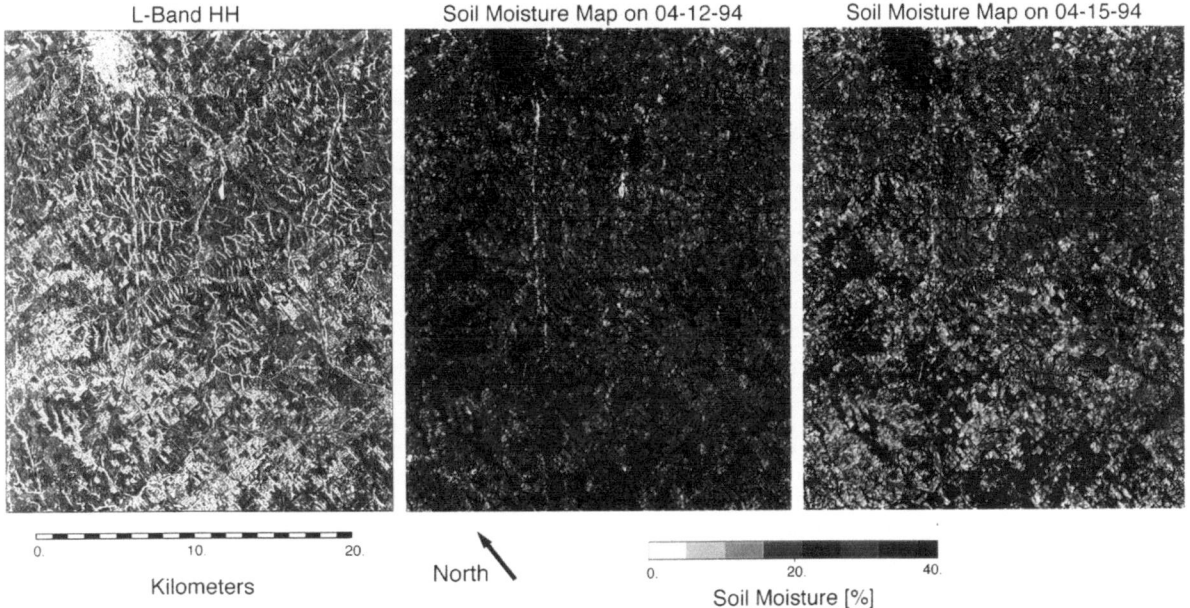

FIGURE 6.4 L-Band hh-polarized image and two soil-moisture maps derived from Shuttle Imaging Radar-C data for April 12 and 15, 1994. The first day was extremely wet, and the second day was drier (Ulaby et al., 1996).

The results of the SIR-C and other similar investigations, which are based only on L-band SAR observations, provide strong support for radar as a soil-moisture mapper. The bare-soil algorithms used in the generation of the soil-moisture maps ignore the presence of vegetation cover, which can lead to significant soil-moisture estimation errors for soil surfaces covered with vegetation whose biomass exceeds 0.5 kg/m^2. For vegetation-covered soils, it is necessary to use two-frequency channels, as discussed below. An alternative approach, similar to that found to be useful in passive microwave radiometry of vegetated soils, would be to develop vegetation correction factors based on fusion with electro-optical data. Multispectral fusion approaches, designed to address this challenge, have not been seriously developed and tested.

Vegetation-Covered Surfaces

Using the combination of L- and C-band multipolarized SAR data, it is possible to classify terrain with an accuracy exceeding 95 percent for simple land-use/cover classes (Dobson et al., 1995; Pierce et al., 1994), which is on a par with the accuracy of ground-truth data. These land-use/cover classes include six basic types: (1) tall vegetation (trees), (2) short vegetation (cultural vegetation, grasses, rangeland), (3) urban areas, (4) bare surfaces, (5) water, and (6) flooded lands and swamps. The same SAR data are then applied to classify the short-vegetation pixels into a finer level, according to vegetation structure from a radar perspective. For each structural class, a soil-moisture algorithm can be applied to estimate both the soil-moisture content and the vegetation biomass. An example comparing SAR-predicted values of these quantities versus those measured in situ is given in Figure 6.5 for a soybean canopy (Ulaby, 1998).

Complementarity of Active and Passive Microwave Observations

Table 6.1 summarizes the sensor configurations appropriate for passive and active microwave techniques for mapping the distribution of soil moisture. To produce soil moisture maps with 10-km pixels from a satellite altitude of 600 km, a passive microwave radiometer would need an antenna 15 to 20 m in diameter or a rectangular aperture of comparable dimensions.

Two competing technologies have been proposed in the past 3 years, one involving an inflatable structure and the other using a thinned-array configuration. The objective of both designs is to develop a lightweight antenna with a large effective aperture. Each of the two aperture approaches offers some unique advantages; it is too early to make a clear-cut choice between them. The advantage of the inflatable antenna is that it can be made to scan in a conical format at an approximately constant incidence angle, thereby simplifying the soil-moisture inversion algorithm, but inflatable technology has not yet been tested for a 20-m-diameter antenna. The other approach, relying on the application of signal correlation to a large number of receivers, each connected to a linear antenna, ends up "thinning down" the aperture by a factor of about 5. Its proposed scan mode, however, is in the plane orthogonal to the direction of flight, encompassing a wide incidence angle range extending between nadir and 45°. Hence, a separate correction has to be applied to each beam position.

TABLE 6.1 Sensor Configurations for Mapping Soil Moisture with Active and Passive Microwave Techniques

Parameter	Passive Microwave	Active Microwave (SAR)
Frequency band	Bare soil: L-band Vegetation-ground: L-band + NDVI correction	Bare soil: L-band[a] Vegetation covered: L + C
Polarization(s)	Horizontal	Dual or polarimetric
Antenna size	20 m × 20 m	10 m × 0.5 m
Resolution from 600 km orbit	10 km	100 m
Swath width	1,000 km	500 km

[a]A longer wavelength, such as P-band, will also be suitable.

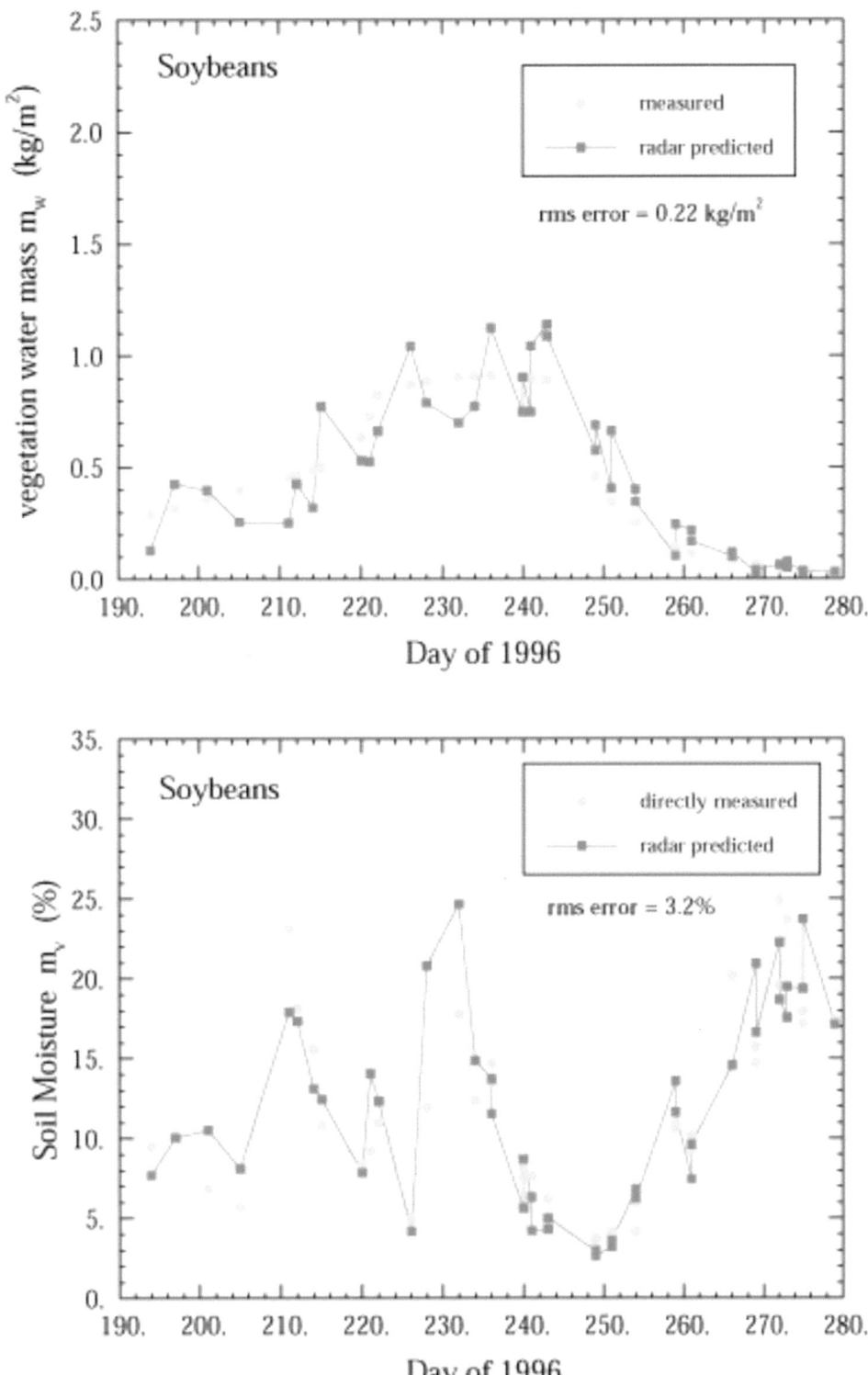

FIGURE 6.5 Measured versus predicted vegetation water mass (m_w) and soil moisture (m_v) (Ulaby, 1998).

To address the needs of applications requiring finer-resolution soil-moisture information, including small-basin hydrology, ecology, and agricultural concerns, a moderate-resolution (100 to 300 m) SAR can be used. When operated at such a moderate resolution, the SAR can generate images with swaths of about 500 km while maintaining the data rate at a level acceptable for transmission. A dual-polarized L-band (or longer wavelength) SAR can generate soil-moisture maps for bare-soil surfaces (which by the previous definition include surfaces with 10-cm or shorter vegetation covers) with good accuracy. To handle vegetation-covered surfaces, an additional frequency is needed, preferably the C-band. The active and passive soil-moisture microwave mappers do not have to operate from the same platform, but there are some decided advantages to operating them together (Figure 6.6).

OBSERVING STRATEGY OF CURRENT AND FUTURE SATELLITE SENSORS

Sensors Currently in Use

Of the passive microwave sensors currently flying aboard orbiting satellites, none operates at the L-band (l = 21 cm) and none is suitable for soil-moisture mapping. In spite of this, some attempts have been made to use Special Sensor Microwave/Imager (SSMI) data that are weakly correlated to soil moisture. Currently, no L-band radiometer systems are on any approved list for development and flight on a future U.S. mission or as part of any other space program. NASA solicited recommendations for post-2002 missions to support the Earth Science Enterprise in 1998. Both the interdisciplinary panel and the disciplinary panel in atmospheric chemistry and hydrology strongly supported a soil moisture mission. The group gave a high priority to a recommendation for use of low-frequency microwave radiometry techniques as an exploratory mission.

The SAR picture is less bleak; Japan operates an L-band SAR and ESA operates a C-band SAR, as does Canada (but with different polarization). All three are single-polarization sensors incapable of mapping soil moisture reliably. ESA plans to fly a dual-polarized C-band configuration on ENVISAT, which can be used to map soil moisture of bare-soil surfaces, but under a moderately limited range of conditions because the C-band is more sensitive to vegetation cover. Also, under current plans, the Japanese space agency is considering launching a dual-polarized L-band SAR in 2001. Such a system will be suitable for mapping soil moisture of bare-soil surfaces.

Future Sensors

The most promising future sensor for demonstrating the mapping of soil moisture is LightSAR because it will provide the needed combination of bands (both L and C) and polarizations (all states) to generate maps of terrain classes, to classify vegetation by structural class, and to map soil moisture content. It is important to note, however, that because soil-moisture mapping was not one of the key objectives of the LightSAR design, it was configured to generate high-resolution images (1 to 30 m) with narrow swaths (20 to 50 km) rather than moderate-resolution images with wide swaths. It will be possible to demonstrate that LightSAR data can indeed be used for the generation of soil-moisture maps, but their limited swath width (and correspondingly long revisit interval) will make them useful in fewer applications.

CALIBRATION AND VALIDATION

The design and operation of any satellite system for measuring soil moisture should consider sensor calibration and product validation carefully. Calibration and validation must be an integral part of mission planning. Relative and absolute calibration must be monitored over the instrument's lifespan to enable time series analyses of derived fields. Metadata must be sufficient to permit reprocessing of archived data with improved algorithms as they become available. Validation of estimates is especially challenging because of the natural spatial variability of soil moisture. Soil moisture is often measured in situ to support agricultural research, and such measurements may be made weekly or monthly on a long-term basis. Common methods include gravimetric measurement, time

SOIL MOISTURE

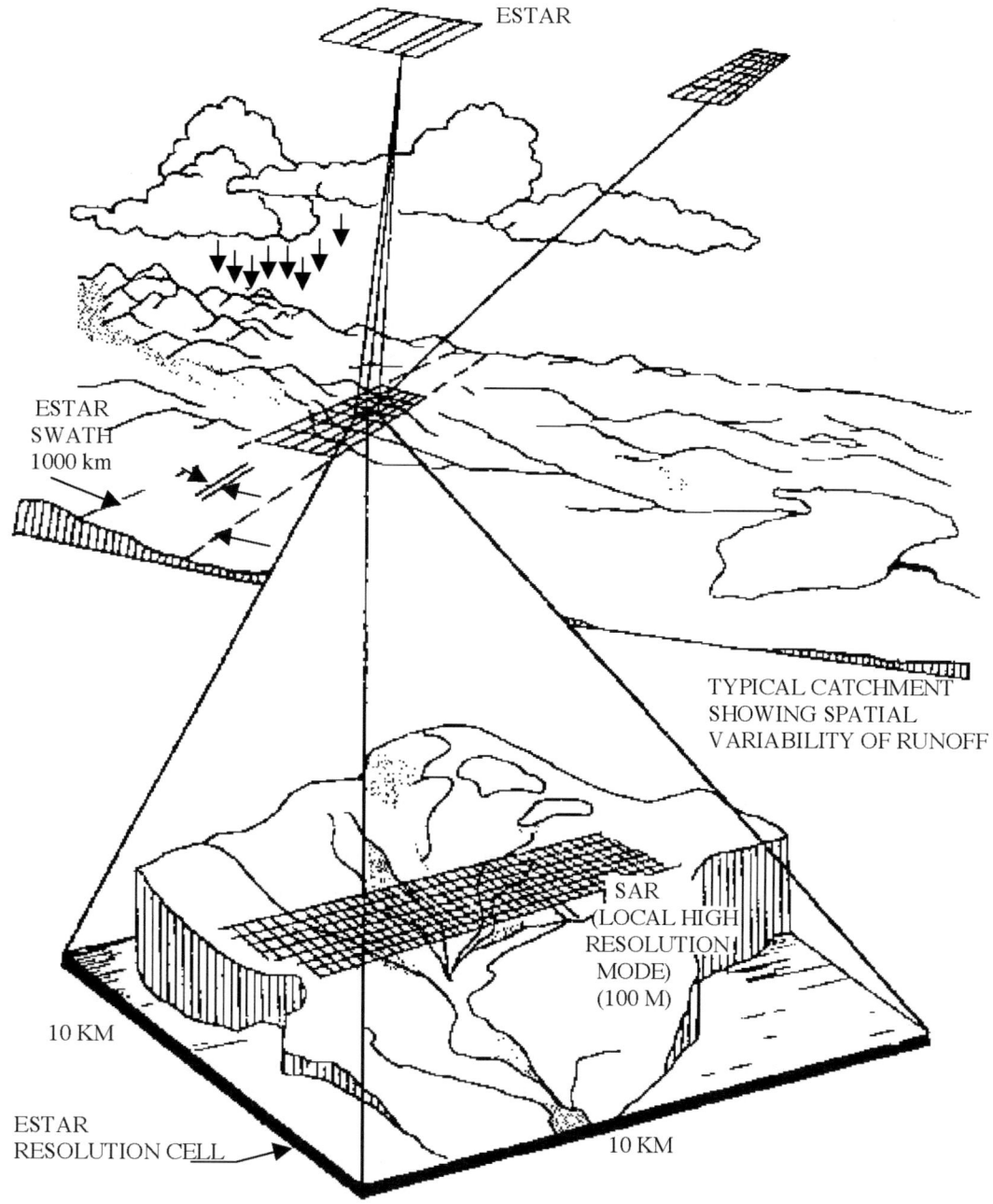

FIGURE 6.6 Schematic showing synergistic operation of synthetic aperture radar (SAR) in the high-resolution mode and electronically scanned thinned array radiometer (ESTAR).

domain reflectivity (TDR), and neutron probes, generally conducted at one location or over a sparse network of locations and at several depths. Usually the vertical profile is well defined, but spatial sampling is not sufficient to detect or characterize gradients in the horizontal dimension. A high degree of spatial variability in soil properties such as moisture and hydraulic conductivity has been observed experimentally (Hawley et al., 1982; Rogowski, 1972). Thus, coarse-resolution estimates derived from satellite observations cannot be readily validated using such data. Measurements of local variability of soil moisture within agricultural fields often show standard deviations of about 2 to 3 percent (Bell et al., 1980). Hence, the local uncertainty even of intensive point sampling is comparable to the demonstrated uncertainties of the remotely sensed estimates. Consequently, a network of validation sites will have to be instrumented with calibrated arrays to record both the time-variant mean soil moisture and higher statistical moments. External validation cannot be accomplished using sparse point samples because of the extreme spatial and temporal variability of soil moisture; a network of well-calibrated arrays is required. The network should be distributed over a variety of land-cover types. Vegetation attributes such as type, leaf area index, and biomass should be monitored as well, to calibrate and validate vegetation effects on soil-moisture retrieval algorithms.

EVOLUTION STRATEGY

Considering the importance of soil moisture to climate physics and to the fields of hydrology, ecology, and agriculture, the committee strongly recommends that NASA provide the resources necessary to develop and launch soil-moisture mapping systems in the near future. A development strategy should be defined in conjunction with other relevant agencies such as NOAA and the U.S. Department of Agriculture (USDA). NASA has initiated strategic planning for a soil-moisture mission as part of its post-2002 Earth Science Enterprise planning and within the NASA Hydrology Program. The strategy has to define specific measurement requirements for various scientific and operational objectives with respect to land-cover type and spatial and temporal scales. The relative priorities of requirements can guide the development of specific technological capabilities.

The effects of vegetation are central to the discussion of microwave techniques for estimating soil moisture. A misperception that radiometry is less sensitive than radar to vegetation effects is largely the consequence of very few experimental efforts using radiometry outside arid to semi-arid landscapes, where vegetation cover is scant. Experimental work conducted at Texas A&M and the University of Kansas in the 1970s using truck-mounted radiometer and scatterometer observations of agriculture crops showed similar reductions in sensitivity to soil moisture for both the active and passive microwave techniques. Recent theoretical treatment of the issue also confirms that the masking effects of vegetation are similar for radiometers and radar in most instances (Du et al., 2000). Beyond some threshold of plant biomass, both active and passive measurement techniques must rely on multispectral fusion to estimate the type and quantity of overlying vegetation and to correct for the effect on emission or scattering. These correction factors can be derived from fusion with electro-optical data or microwave data with wavelength less than approximately 5 cm.

Current space-based assets are inadequate for absolute measurement of soil moisture. The next generation of orbital multipolarized SAR now under construction, ESA's ENVISAT and NASDA's PALSAR, will be capable of mapping near-surface soil moisture where vegetation cover is less than approximately 1 kg/m^2. Coarse-resolution measurement of soil moisture can be accomplished with an L-band radiometer for bare soils. Extending measurement capability to vegetated (but nonforested) conditions will require the development of a multifrequency SAR or further development of multispectral fusion techniques. The ultimate success of any space-based measurement capability will be determined through calibration and validation with in situ measurements. It is critical that an in situ soil-moisture measurement strategy be developed to complement the remote-sensing strategy.

Information on the spatial distribution and temporal variation of the near-surface soil moisture content is needed over a wide range of spatial scales, from 10 m for some agricultural applications to 10 km or more for certain climate models. Coarse-resolution (10 km), wide-swath (1,000 km) soil-moisture maps are best provided by a horizontally polarized L-band radiometer using either a large inflatable antenna or a thinned array configuration. Moderate-resolution (100 m to 1 km) soil-moisture data can be provided at a swath width of 500 km by a dual-frequency (L- and C-bands), dual-polarized SAR.

REFERENCES

Bell, K.R., B.J. Blanchard, T.J. Schmugge, and M.W. Witczak. 1980. Analysis of surface moisture variations within large-field sites. Water Resour. Res. 16(4): 776-781.

Betts, A.K., J.H. Ball, A.C. Baljaars, M.J. Miller, and P. Viterbo. 1994. Coupling between land surface, boundary layer parameterizations and rainfall on local and regional scales: Lessons from the wet summer of 1993. Fifth Conference on Global Change Studies, American Meteorological Society 74th Annual Meeting, Nashville, Tenn., Jan. 23-28.

Chang, J.T., and P.J. Wetzel. 1991. Effects of spatial variations of soil moisture and vegetation on the evolution of a prestorm environment: A numerical case study. Mon. Weather Rev. 119(6): 1368-1390.

Choudhury, B.J., T.J. Schmugge, R.W. Newton, and A. Change. 1978. Effect of surface roughness on microwave emission of soils. J. Geophys. Res. 84: 5699-5706.

Delworth, T.L., and S. Manabe. 1988. The influence of potential evaporation on the variabilities of the simulated soil wetness and climate. J. Climate 1: 523-547.

Dobson, M.C., F.T. Ulaby, M. Hallikainen, and M. El-Rayes. 1985. Microwave dielectric behavior of wet soil, Part II: Dielectric mixing models. IEEE Trans. Geosci. Remote Sensing. 23(1): 35-46.

Dobson, M.C., F.T. Ulaby, and L.E. Pierce. 1995. Land-cover classification and estimation of terrain attributes using synthetic aperture radar. Remote Sensing Environ. ISLSCP Special Issue. 51: 199-214.

Du, Y., F.T. Ulaby, and M.C. Dobson. 2000. Sensitivity to soil moisture by active and passive microwave sensors. IEEE Trans. Geosci. Remote Sensing, in press.

Dubois, P.C., J. van Zyl, and T. Engman. 1995. Measuring soil moisture with imaging radars. IEEE Trans. Geosci. Remote Sensing 33: 915-926.

Evans, D.L., J. Apel, R. Arvidson, R. Bindschadler, F. Carsey, J. Dozier, K. Jezek, E. Kasischke, F. Li, J. Melack, B. Minster, P. Mouginis-Mark, and J. van Zyl. 1995. Spaceborne Synthetic Aperture Radar: Current Status and Future Directions. NASA Technical Memorandum 4679.

Fast, J.D., and M.D. McCorcle. 1991. The effect of heterogeneous soil moisture on a summer baroclinic circulation in the central United States. Mon. Weather Rev. 199(9): 2140-2167.

Food and Agriculture Organization (FAO), United Nations. 1995. World Agriculture: Towards 2010. An FAO Study. Nikos Alexandratos (ed.). New York: John Wiley & Sons.

Hallikainen, M., F.T. Ulaby, M.C. Dobson, M. El-Rayes, and L.K. Wu. 1985. Microwave dielectric behavior of wet soil, part I: Empirical models and experimental observations. IEEE Trans. Geosci. Remote Sensing. 23(1): 25-34.

Hawley, M.E., R.H. McCuen, and T.J. Jackson. 1982. Volume-accuracy relationship in soil moisture sampling. J. Irrig. Drain. Div. Am. Soc. Civ. Eng. 108(IR1): IR1-IR11.

Jackson, T.J. 1993. Measuring surface soil moisture using passive microwave remote sensing. Hydrol. Processes 7: 139-152.

Jackson, T.J., and D.E. Le Vine. 1996. Mapping surface soil moisture using an aircraft-based passive microwave instrument: algorithm and example. J. Hydrol. 184: 85-99.

Jackson, T.J., and T.J. Schmugge. 1991. Vegetation effects on the microwave emission of soils. Remote Sensing Environ. 36: 203-212.

Jackson, T.J., D.M. Le Vine, A.J. Griffis, D.C. Goodrich, T.J. Schmugge, C.T. Swift, and P.E. O'Neill. 1993. Soil moisture and rainfall estimation over a semiarid environment with the ESTAR microwave radiometer. IEEE Trans. Geosci. Remote Sensing 31: 836-841.

Jackson, T.J., D.M.Le Vine, C.T. Swift, T.J. Schmugge, and F.B. Schiebe. 1995. Large scale mapping of soil moisture using the ESTAR passive microwave radiometer in Washita '92. Remote Sensing Environ. 53: 27-37.

National Oceanic and Atmospheric Administration (NOAA). 1997. Climate Measurement Requirements for the National Polar-orbiting Operational Environmental Satellite System (NPOESS), Workshop Report, Herbert Jacobowitz (ed.), Office of Research Applications, NESDIS-NOAA, February. 77 pp.

Lanicci, J.M., T.N. Carlson, and T.T. Warner. 1987. Sensitivity of the Great Plains severe-storm environment to soil moisture distribution. Mon. Weather Rev. 115(11): 2660-2673.

Le Vine, D.M., M. Kao, A.B. Tanner, C.T. Swift, and A. Griffis. 1990. Initial results in the development of a synthetic aperture microwave radiometer. IEEE Trans. Geosci. Remote Sensing 28: 614-619.

National Oceanic and Atmospheric Administration (NOAA). 1997. Weather and Climate Observing Systems: An Investment for America in the 21st Century. Boulder, Colo.: University Center for Atmospheric Research.

Oh, Y., K. Sarabandi, and F.T. Ulaby. 1992. An empirical model and an inversion technique for radar scattering from bare soil surfaces. IEEE Trans. Geosci. Remote Sensing 30: 370-382.

Pierce, L.E., F.T. Ulaby, K. Sarabandi, and M.C. Dobson. 1994. Knowledge-based classification of polarimetric SAR images. IEEE Trans. Geosci. Remote Sensing 32: 1081-1086.

Rind, D. 1982. The influence of ground moisture conditions in North America on summer climate as modeled in the GISS GCN. Mon. Weather Rev. 110(5): 1487-1494.

Rogowski, A.S. 1972. Watershed physics: soil variability criteria. Water Resour. Res. 8(4): 1015-1023.

Shukla, J., and Y. Mintz. 1982. The influence of land-surface evapotranspiration on Earth's climate. Science 215: 1498-1501.

Ulaby, F.T. 1998. SAR algorithm for mapping soil moisture and vegetation biomass. Soil Hydrology Meeting, Baltimore, Md., March.

Ulaby, F.T., R.K. Moore, and A.K. Fung. 1986. Microwave Remote Sensing: Active and Passive. Vol. III. Dedham, Mass.: Artech House.

Ulaby, F.T., P.C. Dubois, and J. van Zyl. 1996. Radar mapping of surface soil moisture. J. Hydrol. 184: 57-84.

Walker, J.M., and P.R. Rowntree. 1977. The effect of soil moisture on circulation and rainfall in a tropical model. Q.J.R. Meteorol. Soc. 103: 29-46.

7

Aerosols

INTRODUCTION

Aerosols are suspensions of solid or liquid particles in a gas. The particles that compose the atmospheric aerosol range in size from nanometers (in the case of large clusters of molecules) to tens of micrometers (in the case of wind-driven sand). Some aerosols (e.g., sea salt and terpene haze) occur naturally and some (e.g., smoke) are man-made (anthropogenic). Aerosols represent one of the greatest uncertainties in climate modeling, and they can affect climate in two ways: (1) by absorbing or scattering both shortwave and longwave radiation, they alter the radiative properties of the atmosphere (the direct effect) and (2) by serving as cloud condensation nuclei (CCN), they play a critical role in the cloud formation process, changing the radiative properties of the clouds and, possibly, their physical structure and precipitation (the indirect effect).

Anthropogenic aerosols could therefore to some extent offset the global warming due to greenhouse gases (GHGs). To fully understand how aerosols affect climate, their characteristics (composition, size distribution, and optical properties) must be measured on a global scale. Aerosols reside mainly in the two lowest layers of the atmosphere, the troposphere and the stratosphere.

BASIC SCIENCE ISSUES

Tropospheric Aerosols

Tropospheric aerosols, a substantial proportion of which are anthropogenic, form a much more complex system than aerosols in the stratosphere. The aerosols may be surface-derived from both land and ocean or formed in the atmosphere as a result of gas-to-particle conversion or cloud cycling. Once in the atmosphere, they may be transported away from their place of origin, sometimes over great distances. They may be removed from the atmosphere by both dry processes (sedimentation) and wet (rainout). In the troposphere the aerosol concentration generally decreases with altitude, reaching its lowest values in the upper troposphere.

The composition of tropospheric aerosols is variable; mixtures are formed both internally (within a single particle) and externally (between particles). For the purpose of behavioral description and modeling, tropospheric aerosols are commonly classified according to their composition and source, because they vary significantly in concentration and composition by region (source). They have horizontal spatial scales ranging from about 1 km to a few thousand kilometers. The highly visible haze that persists in all the industrialized regions of the world is composed mainly of sulfates and organic compounds from emissions of sulfur dioxide (SO_2); organic gases (e.g., terpenes); and organic matter and soot (carbon black) from biomass burning. The flux of SO_2 emissions has increased exponentially over the past century to 65 to 80 teragrams (1 Tg = 10^{12} g) per year, mainly from the smelting of metal ores and the burning of fossil fuels, which has led to increased emissions of greenhouse gases, aerosol particles, and aerosol precursor gases as well.

The ocean is a significant source of natural tropospheric aerosols. Air-sea exchange of particulate matter contributes to the global cycles of carbon, nitrogen, and sulfur aerosols (an example of the last-mentioned is the dimethyl sulfide (DMS) produced by phytoplankton). Ocean water and sea salt are transferred to the atmosphere through air bubbles at the sea surface. As the water evaporates, the salt is left suspended in the atmosphere. Haywood et al. (1999) suggest that naturally occurring sea salt is the leading aerosol contributor to the global-mean clear-sky radiation balance over oceans. Other significant sources of natural tropospheric aerosols are volcanic eruptions and windblown dust from arid and semiarid regions.

While the direct and indirect radiative effects of sulfates are important, other tropospheric aerosols may contribute significantly to the global radiative balance. Among these are carbonaceous compounds and mineral dust. Carbonaceous compounds are present in the atmosphere in the form of elemental carbon (EC) or organic carbon (OC). A significant portion of ambient EC is soot directly emitted as a product of incomplete fossil fuel combustion. An important property of EC is its large share of the imaginary part of the refractive index at visible wavelengths. This property makes it a very good absorber of shortwave radiation and could decrease the single-scattering albedo of an aerosol to below the critical point, causing the aerosol to have a net heating effect instead of a cooling effect. Simple radiative-transfer calculations using a box model show that EC/SO_4 ratios of 0.05 and 0.10 result in a positive forcing (heating) of +0.03 and +0.34 Wm^{-2}, respectively. In comparison, the direct sulfate forcing has been estimated at -0.43 Wm^{-2} for the northern hemisphere.

OC is either directly emitted (primary OC) or formed in the atmosphere (secondary OC) by the condensation of volatile organic carbons (VOCs). The largest sources of anthropogenic organic carbon include biomass burning, dust from paved roads, industrial emissions, and combustion for domestic purposes (e.g., cooking of food and burning of wood in stoves and fireplaces). Because a number of secondary and primary forms of OC are hygroscopic and have size distributions and optical properties similar to those of sulfate particles, these OC particles are likely to force the climate as much as, or even more than, sulfate particles.

The biggest obstacle to determining the effect of these particles for use in climate models is the lack of well-defined, spectrally resolved refractive indices for determining fundamental optical properties (single-scattering albedo, asymmetry factor, and extinction efficiency). The refractive indices of these particles have not been well characterized because ambient OC is made up of more than 300 different compounds. As a result, the composition is highly variable from particle to particle depending on location, source, and meteorological conditions. Based on our current knowledge, any estimate of a set of optical properties for all OCs would entail great uncertainty. Penner et al. (1994) estimate up to -0.8 Wm^{-2} (direct and indirect) forcing as a result of anthropogenic organic aerosols produced by biomass burning. There is evidence that organic aerosols play a key role in cloud nucleation and thus are responsible for a significant share of cloud albedo enhancement in regions affected by anthropogenic pollutants. Based on measurements made on El Yunque peak in eastern Puerto Rico, 37 percent (by number) of the total CCN were found to be sulfate particles and the remaining 63 percent were OC. Some OC particles are strongly hydrophilic and readily act as CCN. Others may be intrinsically inactive as nuclei but become active by the condensation of a thin coating of sulfuric acid.

Mineral dust absorbs and scatters solar radiation and absorbs terrestrial (infrared) radiation. Although there has been considerable interest in sulfate aerosols over the last two decades, our knowledge of the distributions, global burdens, and effects on climate change of elemental carbon, organic carbon, and mineral dust is meager compared to our knowledge of sulfate aerosols.

Although tropospheric aerosols are chemically complex and may be strongly influenced by local emissions, one persistent feature, worldwide, is the strong presence of sulfate. It is difficult to calculate the effect of tropospheric aerosols on Earth's climate because data are lacking for many places around the world and there is no clear understanding of the processes that link gas emissions with particle formation and growth. All the estimates to date of the global effects of anthropogenic aerosols have been based solely on coupled radiative and chemical transport models.

Stratospheric Aerosols

Stratospheric aerosol particles are composed mostly of sulfuric acid (H_2SO_4) and water (H_2O) droplets less than a micron in diameter. They are present globally between the tropopause and about 30 km, undergo seasonal variations, and are greatly influenced by large volcanic eruptions. During volcanically quiescent periods, the vertical distribution of the stratospheric aerosol particles relative to tropopause height is very similar at all latitudes, with mass mixing ratios and number densities on the order of 1 ppbm and 10 particles cm^{-3}, respectively. The predominant source of stratospheric sulfate aerosols is strong, sulfur-rich volcanic eruptions, which are by nature highly intermittent and unpredictable. The flux of volcanic sulfur averaged over the last 200 years has been estimated at about 1 Tg yr^{-1}, with lower and upper bounds of 0.3 and 3 Tg yr^{-1} (Pyle et al., 1996). A minimum flux of 0.5 to 1 Tg yr^{-1} for the past 9,000 years has been derived from ice core sulfate data. The volcanic input into the stratosphere has been unusually high during the past 15 years, with the occurrence of two relatively large sulfur-rich eruptions: 3.5 Tg from El Chichon (1982) and 9 Tg from Mt. Pinatubo (1991).

Carbonyl sulfide (OCS) oxidation is believed to be the main nonvolcanic source of stratospheric sulfur (Crutzen, 1976). Recent estimates of this source range from 0.03 Tg yr^{-1} (Chin and Davis, 1995) to 0.049 Tg yr^{-1} (Weisenstein et al., 1997). Although most OCS sources are natural, there are some indications that anthropogenic emissions may be substantial and increasing (Zander et al., 1988; Khalil and Rasmussen, 1984; Hofmann, 1990). However, historical data on industrial releases suggest that anthropogenic emissions of OCS and its precursor, carbon disulfide (CS_2), were relatively constant between 1977 and 1992 (Chin and Davis, 1993). Furthermore, no statistically significant trend in lower stratospheric OCS was inferred from spaceborne observations made in 1985 and 1994 (Rinsland et al., 1996).

The main mechanism for removal of stratospheric aerosols is a combination of gravitational settling and stratospheric-tropospheric exchange. Typically, about one-third is removed in a year.

Another major class of stratospheric particles is the polar stratospheric clouds (PSCs) (McCormick et al., 1982) observed in cold regions of the lower polar stratosphere, primarily during winter. Based on their optical properties, PSCs have been further divided into distinct subclasses: type 1 PSCs are thought to be relatively small and rich in nitric acid (HNO_3), and type 2 PSCs are larger, primarily H_2O ice particles. Typical mass mixing ratios for type 1 and type 2 PSCs are 10 and 1,000 ppbm, respectively (Drdla, 1996).

Since the discovery of the stratospheric aerosol layer in 1957 (Junge et al., 1961), there has been much speculation about the stability of the layer and the background source of the H_2SO_4 that is the primary component of the aerosol. The measurements by Junge et al. (1961) were made at the end of a long period free of volcanic eruptions (Stothers, 1996) but were not extensive enough to establish a baseline. There are four periods in the modern (post-1970) measurement era during which the influence of volcanic eruptions has been at a minimum: 1974, 1979, 1989 to early 1991, and the present. Many studies have focused on these data periods in an attempt to clarify the processes that sustain the background nonvolcanic stratospheric aerosol layer and to explain the cause(s) for changes observed from one period of minimum volcanic activity to another.

Both tropospheric and stratospheric aerosols play an important role in global climate change. Natural variations of aerosols, especially those due to episodic eruptions of large volcanoes, are recognized as a significant forcer of climate; that is, they alter the planetary radiation balance and thus tend to cause global temperature change. In addition, there are several ways in which humans are altering atmospheric aerosols and thereby possibly affecting climate. The concern here is with radiative forcing of climate due to changing aerosols, both direct and indirect. These climate forcings are not well determined, especially the forcing by anthropogenic aerosols. Findings from recent studies suggest that anthropogenic aerosols, primarily sulfates, organics, and carbon black, induce a significant radiative forcing opposite in sign to radiative forcing by anthropogenic GHGs. However, the GHG and anthropogenic aerosol forcings have very different spatial and temporal scales. In industrialized areas, aerosol forcing can be much larger than GHG forcing. According to the Intergovernmental Panel on Climate Change (IPCC), the negative direct effect, worldwide, of tropospheric aerosol forcing is about

20 percent of the effect of forcing from GHGs (with an uncertainty range of 0 to 40 percent); the indirect effect of aerosols is highly uncertain but has been estimated to be even larger than their direct effect. GHG forcing exists during the day and at night, whether it is clear or cloudy, and attains a maximum in the hottest, driest locations on Earth. Anthropogenic aerosol forcing exists mainly during the day, attains a maximum in clear conditions, and—because the aerosols have relatively short residence times—is concentrated near aerosol source regions. Negative forcings as large as 40 to 60 Wm^{-2} have been reported to occur at midday.

The National Research Council (NRC) report *Aerosol Radiative Forcing and Climate Change* (1996) recommended that the uncertainties in calculated aerosol forcing at the top of the atmosphere be reduced to within ±15 percent, both globally and locally. Locally, this would imply an uncertainty in forcing of less than 1.5 Wm^{-2}. Estimates of climate forcings by GHGs, aerosols, and other forcers have been reported in a number of publications (e.g., IPCC, 1995). Figure 7.1 shows the climate forcings by several different agents (Hansen et al., 1998). GHG forcing is estimated at 2.3±0.25 Wm^{-2}, while tropospheric aerosols are estimated to force climate on a global scale −0.4±0.3 Wm^{-2}. Indirect effects are estimated at −0.5 to −2 Wm^{-2}. Because of the great spatial variability in tropospheric aerosol concentrations that results from the aerosols' short lifetimes, there are many regions, principally near major industrial areas, where aerosol negative forcing exceeds the greenhouse positive forcing (e.g., Charlson et al., 1992).

Recent studies have shown that (1) aerosol effects appear to be present in the global and regional 20th-century temperature record and (2) the inclusion of aerosol effects in numerical model predictions improves agreement with observed temperatures, in both timing and spatial patterns (Karl et al., 1995). Volcanic aerosol forcing, albeit episodic, is estimated at 0.2 to −0.5 Wm^{-2}. About 1 year after the Mt. Pinatubo eruption, for example, forcing was estimated at about −3 Wm^{-2}, which is greater than GHG forcing. Also shown in Figure 7.1 is an estimate of negative forcing by lower stratospheric ozone depletion of about −0.2±0.1 Wm^{-2}. Mid- to lower-stratospheric ozone depletion is thought to be caused primarily by heterogeneous reactions on stratospheric volcanic aerosol particles and therefore represents an indirect effect of aerosols in addition to changes in clouds, mentioned earlier. Heterogeneous reactions on the surfaces of PSCs and volcanic aerosols have been shown to be the key to understanding ozone depletion (Solomon et al., 1986, 1996; McElroy et al. 1986).

As can be seen from Figure 7.1, aerosols have been one of the greatest sources of uncertainty in the interpretation of climate change during the past century and in the projection of future climate change. In addition to their radiative effects and their effects on stratospheric chemistry, aerosols are also important to tropospheric chemistry, air quality, acid deposition, visibility, and cloud and precipitation processes.

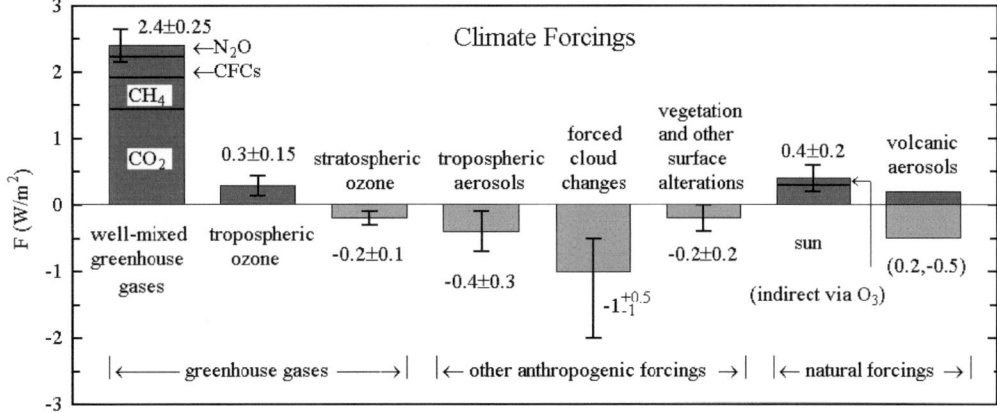

FIGURE 7.1 Estimates of climate forcing by greenhouse gases, other anthropogenic forcers, and natural forcers (Hansen et al., 1998).

Some attention has also recently focused on carbon black soot aerosols in the stratosphere, again because of their potential heterogeneous chemical reactivity. Measurements (Blake and Kato, 1995; Pueschel, 1996) and model calculations (Bekki, 1995) indicate that aircraft emissions are the most important source of soot in the stratosphere. Maximum soot concentrations on the order of 1 ng m^{-3} are found at northern midlatitudes around aircraft cruise altitudes (about 10 km). Most of the soot particles end up embedded in H_2SO_4/H_2O solutions via coagulation with H_2SO_4/H_2O aerosols and, possibly, the condensation of gaseous H_2SO_4. In the troposphere, black carbon aerosol has been detected near industrial regions and even in regions considered remote from anthropogenic sources. In contrast to sulfate and organic aerosol, black carbon aerosol is a strong absorber of solar radiation and can lead to localized warming rather than cooling.

OBSERVING STRATEGY

During the past 20 to 25 years, characteristics of stratospheric aerosol have become much better known as a result of in situ measurements from balloons and aircraft as well as remote sensing by lidar and satellites. Model studies have paralleled the observational program. The satellite measurements by the Stratospheric Aerosol Measurement (SAM) II, Satellite Aerosol and Gas Experiment (SAGE) I, and SAGE II have given the most complete global picture of these characteristics. Ground-based lidar networks, although mainly in the northern mid- to high latitudes, have also contributed to this more global view and the Polar Ozone and Aerosol Measurement (POAM) II and Halogen Occulation Experiment (HALOE) instruments have recently added to this stratospheric aerosol data set (Bevilacqua, 1997; Russell et al., 1993). However, the SAM II/SAGE series of measurements is the only global source of upper tropospheric aerosol information (Kent et al., 1988) and the longest global stratospheric aerosol database available.

Whereas the stratospheric aerosol is more homogeneous in composition and controlled primarily by episodic volcanic eruptions, the tropospheric aerosol is more heterogeneous in composition and location and is controlled by the myriad of aerosol sources in each region. Global data on tropospheric aerosol are sparse. Tropospheric aerosols may be observed from space by measuring solar radiation scattered back from the atmosphere (Kaufman, 1995). Such measurements reveal most clearly the larger and lower-altitude aerosol concentrations such as desert dust clouds and pollution episodes. Converting these measurements to quantitative estimates of aerosol concentration is made difficult by the problem of separating the aerosol signature from that of the background and by the presence of multiple scattering. Because of these factors, tropospheric aerosol studies from space are largely confined to studies over the oceans (Griggs, 1975; Rao et al., 1989; Durkee et al., 1991), although several techniques have been used to retrieve aerosol properties over land (Kaufman, 1995; Holben et al., 1992). The SAM/SAGE series of satellites, using solar occultation with a limb-viewing geometry, has a much greater sensitivity to the presence of aerosols and provides data that are in many ways complementary to those obtained from nadir viewing instruments, although it primarily yields data in the stratosphere and upper troposphere. Nadir viewing instruments lack vertical resolution but possess good horizontal resolution, while SAM/SAGE instruments have good vertical resolution (~1 km) and a horizontal resolution of about 200 km.

Intensive field programs can also provide the data needed to reduce uncertainties and improve the performance of climate prediction models. The International Global Atmospheric Chemistry (IGAC) Program is coordinating four such field programs, including the Tropospheric Aerosol Radiation Forcing Observation Experiment (TARFOX) and the Aerosol Characterization Experiments (ACE-1, ACE-2, and ACE-3) (IGAC, 1996). TARFOX was designed to reduce the uncertainty of aerosol effects on atmospheric radiation by measuring and analyzing aerosol properties and effects on the United States eastern seaboard, where one of the world's largest plumes of urban and industrial haze moves from the continent out over the Atlantic Ocean. The TARFOX intensive field campaign was conducted July 10 through July 31, 1996. It included coordinated measurements from four satellites (GOES-8, NOAA-14, ERS-2, and Landsat), four aircraft (ER-2, C-130, C-131A, and a modified Cessna), land sites, and ships. A variety of aerosol conditions was sampled, ranging from relatively clean, behind frontal passages, to moderately polluted, with aerosol optical depths exceeding 0.5 at mid-visible wavelengths. Gradients of aerosol optical thickness were sampled to aid in separating aerosol effects from other radiative effects and to more tightly constrain closure tests, including those of satellite retrievals. Early results from TARFOX show, among other things, the unexpected importance of carbonaceous compounds and water condensed on aerosols in the U.S. East Coast haze plume, chemical apportionment of the aerosol optical depth, aerosol-induced changes in upwelling and downwelling shortwave radiative fluxes, and generally good agreement between measured flux changes and those calculated from measured aerosol properties (Russell et al., 1999).

The Atmospheric Radiation Measurement (ARM) program is a multilaboratory, interagency program created in 1989 with funding from the U.S. Department of Energy (DOE). The ARM program is part of DOE's effort to resolve scientific uncertainties about global climate change, with a specific focus on improving the performance of general circulation models used for climate research and prediction. These improved models will help scientists better understand the influences of human activities on Earth's climate. The Aerosol Observation System (AOS) is part of the aerosol component of the ARM program. There are three AOS sites with a suite of instruments for characterizing tropospheric aerosols. Each site has a variety of optical particle counters, a single-channel nephelometer, a three-channel nephelometer, a light-absorption photometer, and a condensation nuclei counter, all mounted on a 10-m tower. A Raman lidar and a micropulse lidar are the other instruments for measuring aerosols. The AErosol RObotic NETwork (AERONET) is an optical ground-based aerosol monitoring network set up by NASA and developed to support the earth satellite systems of NASA, the Centre Nationale d'Etudes Spatiales (CNES) of France, and the National Space Development Agency (NASDA) of Japan (Holben et al., 1998). AERONET consists of identical automatic Sun-sky scanning spectral radiometers. Data from this program provide globally distributed near-real-time observations of aerosol spectral optical depths, aerosol size distributions, and precipitable water in diverse aerosol regimes. The main goal of AERONET is to provide algorithm validation of satellite aerosol retrievals, as well as to characterize aerosol properties unavailable from satellite sensors.

The recent, current, and future approved satellite instruments should be sufficient for monitoring most stratospheric aerosol properties important to climate and chemistry. These include three SAGE III instruments, the first scheduled for launch in late 2000 (multiple copies are required to ensure global coverage, since the location at which occultation technique measurements are made depends on spacecraft orbit characteristics). SAGE III is described below in this chapter. The global tropospheric aerosol measurements, however, depend primarily on the success of the satellite-borne MODIS instrument and the ESSP PICASSO-CENA mission and their ability to retrieve aerosol optical depth in Earth's boundary layer (the first few kilometers of altitude). MODIS is the centerpiece of NASA's Earth Observing System (EOS), now aboard Terra (formerly the EOS AM-1 platform), launched on December 18, 1999, and on the EOS-PM platform, to be launched in late 2000. The techniques being used will build on those used to retrieve aerosol optical depth from the AVHRR instrument. With many more wavelength channels, the MODIS science team plans to measure a number of aerosol properties, such as particle size and optical depth. PICASSO-CENA, which will be launched in 2003, uses the lidar technique to achieve high resolution of aerosol characteristics and optical depths. It builds on the proof-of-principle Lidar In-Space Technology Experiment (LITE) flown on the shuttle for 10 days in 1994 (McCormick et al., 1995).

Current Spacecraft Instruments

Table 7.1 summarizes the aerosol measurement capabilities, uncertainties, and vertical resolution of recently flown or now-being-flown instruments designed to collect aerosol information.

SAGE II

NASA's SAGE II has provided dependable stratospheric constituent measurements since October 1984. The SAGE II instrument aboard the Earth Radiation Budget Satellite (ERBS) was launched by the space shuttle Challenger into a 610-km circular orbit with a 57-degree inclination. The SAGE II instrument is a nearly self-calibrating, limb-scanning Sun photometer that measures vertical profiles of aerosol extinction during spacecraft sunrise and sunset. These are measured at four wavelengths: 1.02, 0.525, 0.453, and 0.385 µm. The gaseous absorbers nitrogen dioxide (NO_2), ozone, and water are measured at 0.448 (and 0.453), 0.600, and 0.940 µm, respectively (Mauldin et al., 1985). The SAGE II instrument takes 15 sunset and 15 sunrise measurements each day with a vertical resolution of 1 km. The latitudinal spacing is roughly 0.5 degrees between measurements

TABLE 7.1 Aerosol Measurement Capabilities of Recent or Currently-Being-Flown Spaceborne Instruments

Instrument[a]	Measurement	Inferred Property	Uncertainty (%)	Vertical Resolution
SAGE II	Transmission from the 0.385, 0.453, 0.525, and 1.020 μm wavelengths	Aerosol extinction coefficient, aerosol optical depth, PSC[b] frequency of occurrence	10 to 30 5	1 km
HALOE	Transmission from the 2.45, 3.40, 3.46, and 5.26 μm wavelengths	Aerosol extinction coefficient	15 to 20	2 km
ILAS	Transmission from the 0.780 μm wavelength	Aerosol extinction coefficient, PSC[b] frequency of occurrence	NV[c]	2 km
POLDER	Polarization and directionality from Earth's reflectance	Optical depth over ocean	NV[c]	
POAM II	Transmission from the 0.442, 0.448, and 1.06 μm wavelengths	Aerosol extinction coefficient, PSC[b] frequency of occurrence, PMC[d] frequency of occurrence	20 to 35	1 km
AVHRR	Upwelling radiance from the 0.63 μm wavelength	Aerosol optical depth at the 0.5 mm wavelength over oceans	25	
LITE	Backscatter from the 0.355, 0.532, and 1.064 μm wavelengths	Aerosol backscatter coefficient	10	15 m

[a]Acronyms for instruments are defined in Appendix B.
[b]PSC, polar stratospheric cloud.
[c]NV, not validated.
[d]PMC, polar mesospheric cloud.

depending on latitude, while the longitudinal spacing is approximately 24 degrees. The SAGE II observations cover approximately 75 degrees S to 75 degrees N over a year, providing near-global coverage. The National Centers for Environmental Prediction/Climate Prediction Center (NCEP) supplies the SAGE II database with temperature and pressure data to develop heights at each measurement location. More information can be found on the SAGE II program and early data applications in McCormick (1987), on the instruments in Mauldin et al. (1985), and on the inversion algorithm in Chu et al. (1989). The predecessor instrument, SAGE I, flew aboard the Application Explorer Mission II spacecraft from 1979 through 1981. It had four channels, centered at 0.385, 0.45, 0.6, and 1.02 μm.

HALOE

NASA's HALOE instrument was launched aboard the Upper Atmosphere Research Satellite (UARS) on September 12, 1991, by the space shuttle Discovery into a 585-km, near-circular orbit with a 57-degree inclination. Like SAGE, the HALOE instrument uses the solar occultation technique to measure vertical profiles of O_3, HCl, HF, CH_4, H_2O, NO, and NO_2, extinction due to aerosols, and temperature versus pressure. However, because it uses broadband and gas-filter radiometry methods (Russell et al., 1993) in the spectral range between 2.45 and 10.04 μm, it can provide stratospheric microphysical aerosol information when there are high aerosol loadings, such as occurs during volcanic eruptions (Hervig et al., 1998). Like SAGE II, the HALOE instrument measures approximately 15 sunrise and 15 sunset measurements each day, with similar latitudinal and longitudinal sampling but lower vertical resolution, of approximately 2 km.

ILAS

The Improved Limb Atmospheric Spectrometer (ILAS), another occultation instrument, was launched aboard the Japanese Advanced Earth Observation Satellite (ADEOS) on August 17, 1996, into an 800-km, Sun-synchronous polar orbit with a 98.6-degree inclination (it provides only polar coverage). It measures vertical profiles of O_3, HNO_3, N_2O, NO_2, CH_4, H_2O, CFC-11, CFC-12, N_2O_5, and aerosol extinction in the infrared band between 6.21 and 11.77 µm and temperature, pressure, and aerosol extinction in a visible band centered near 0.78 µm. ILAS measures approximately 14 sunrise and 14 sunset measurements each day with a vertical resolution of 2 km. The sunrise measurements occur entirely at high northern latitudes (55 to 72 degrees N), while the sunset events occur entirely at high southern latitudes (65 to 88 degrees S).

POLDER

The French instrument Polarization and Directionality of the Earth's Reflectances (POLDER) was also launched aboard the ADEOS spacecraft. POLDER measures the polarization, directional, and spectral characteristics of the solar light reflected by aerosols, clouds, oceans and land surfaces. The POLDER instrument is a push-broom-type, wide field-of-view, multiband imaging radiometer and polarimeter designed to measure data in eight narrow spectral bands in the visible and near infrared (0.443, 0.490, 0.565, 0.665, 0.763, 0.765, 0.865, and 0.910 µm). A scientific goal of the POLDER experiment was to determine the physical and optical properties of aerosols so as to classify them and study their variability and cycle (Herman et al., 1997). The ILAS and POLDER instruments operated successfully until June 30, 1997, when the ADEOS satellite malfunctioned (SPARC, 1998). The ILAS II and POLDER instruments are planned as part of the ADEOS II payload. At the time of this writing, the ILAS and POLDER aerosol data are not considered to have been validated.

POAM II

The Naval Research Laboratory's POAM II was launched aboard the French satellite Système Pour l'Observation de la Terre (SPOT) 3 on September 25, 1993, into an 833-km Sun-synchronous polar orbit with a 98.7-degree inclination. The POAM II instrument is also a solar occultation instrument. It is designed as a simpler SAGE II instrument to measure vertical profiles of aerosols, O_3, NO_2, and H_2O in nine channels between approximately 0.35 and 1.06 µm, with 1 km vertical resolution (Glaccum et al., 1996). The instrument performs 14 sunrise and 14 sunset measurements each day; because it is in a Sun-synchronous orbit like ILAS, all sunrise events occur entirely at high northern latitudes (55 to 71 degrees N) and all sunset events occur entirely at high southern latitudes (63 to 88 degrees S). The spacing is approximately equal in longitude (~25.4 degrees) for successive sunrise and sunset events and varies slowly in latitude, with the lowest latitudes measured at the solstices and highest latitudes at the equinoxes. The POAM II database is supplemented by temperature, pressure and potential vorticity for each altitude per measurement location provided by the NCEP. The instrument operated successfully until November 14, 1996, when the SPOT 3 satellite malfunctioned (Bevilacqua, 1997). Aerosol-related products include vertical profiles of polar region aerosols, PSCs, and polar mesospheric clouds (PMCs) (Randall et al., 1996; Fromm et al., 1997; Debrestian et al., 1997). The POAM III instrument was launched in March 1998 aboard the SPOT 4 satellite and is an improved version of POAM II. It has been operational, although the initial aerosol data are considered unvalidated.

AVHRR

The AVHRR instrument flies on the NOAA series of polar-orbiting, Sun-synchronous satellites. The orbital period (time to complete one full orbit around Earth) is approximately 100 minutes, so there are approximately 14 full orbits per day. The nominal altitude of NOAA platforms is about 830 km. The AVHRR is a cross-track scanning system, with a scanning rate of 360 scans per minute. Current AVHRR instruments take data in five narrow-band channels (0.63, 0.83, 3.7, 10.8, and 12 µm). The instantaneous field-of-view for each channel is about 1.4 milliradians, which for a satellite altitude of 830 km leads to a satellite subpoint resolution of approximately 1.1 km. For each scan line (6 per second), the AVHRR takes 2,048 samples per channel that span a viewing angle of ±55 degrees from the nadir (Rao et al., 1989). The aerosol product is optical depth at 0.5 µm wavelength, derived from the 0.63 µm wavelength reflectance data (Stowe et al., 1992).

LITE

The LITE instrument is a three-wavelength (1064, 532, and 256 nm) backscatter lidar developed by NASA and flown on the space shuttle Discovery for 10 days in September 1994. The goals of the LITE mission were to validate key lidar technologies for spaceborne applications, to explore the applications of space lidar, and to gain operational experience that would benefit the development of future systems on free-flying satellite platforms. The performance of the lidar was excellent, as the data gathered presented the first highly detailed global view of the nadir-viewed vertical structure of cloud and aerosol from Earth's surface through the middle stratosphere.

TOMS

The Nimbus-7 and Meteor-3 Total Ozone Mapping Spectrometer (TOMS) instruments, using measured 340-nm and 380-nm radiances, produce daily global maps of ultraviolet (UV)-absorbing aerosols. The same information is currently obtained from the Earth Probe TOMS and was obtained from ADEOS TOMS using the 331-nm and 360-nm wavelength channels. Biomass burning, dust storms, volcanic ash clouds, and even oil fires have been detected by TOMS. Work has been progressing on detecting aerosols that do not absorb UV. Torres et al. (1998) have developed techniques to infer aerosol column optical depths and single-scattering albedo, which includes UV-absorbing aerosols from TOMS measurements in the near-ultraviolet (330 to 400 nm). The main constraint on the ability of these techniques to infer aerosol characteristics is the dependence on external information, such as the type and altitude of the absorbing aerosol present at particular locations and the reflectivity of the surface. Another shortcoming is that they can be applied only in cloud-free conditions.

Future Spacecraft Instruments

SAGE III

The SAGE III series of instruments is part of the EOS program, with the first instrument launch scheduled for December 2000. The SAGE III instrument contains 12 spectral channels over the wavelength region 0.28 to 1.54 µm and is essentially an improved version of its predecessors, SAGE I and II. Whereas previous instruments in the SAGE series used single silicon diodes, the SAGE III instrument uses an 800-pixel charge-coupled device (CCD) linear array detector. The CCD is designed to measure (1) aerosol extinction coefficients centered at wavelengths 0.385, 0.450, 0.521, 0.676, 0.756, 0.869, and 1.0195 µm, (2) absorption features of O_3, NO_2, and H_2O, and (3) both temperature and molecular density profiles from O_2 A-band measurements near 0.760 µm (McCormick et al., 1999). The SAGE III instrument will also have a channel centered near 1.54 µm to improve the size discrimination of larger aerosol particles and to separate cloud and aerosol signals. A unique feature of the SAGE III instrument is the implementation of a lunar occultation mode to additionally measure the active nighttime chemical species NO_3 and chlorine dioxide (OClO). Three SAGE III instruments will enhance and extend the existing database of stratospheric constituents from the SAGE I and II data sets as far back as 1979, when SAGE I was launched.

MODIS

The Moderate-Resolution Imaging Spectroradiometer (MODIS) instrument, to be launched on both the EOS-AM and EOS-PM satellites, measures upwelling scattered radiation in 36 discrete wavelength bands from the visible to the thermal infrared (i.e., 0.4 to 14.5 µm) and will view Earth's entire surface every 1 to 2 days. It uses a conventional imaging radiometer concept, consisting of a cross-track scan mirror and collecting optics, and a set of linear detector arrays. With a spatial resolution of 250 m, 500 m, or 1 km at nadir, MODIS will provide aerosol products in the form of optical thickness, particle size, and mass transport (Esaias et al., 1998; Tanré et al., 1997).

MISR

The Multi-angle Imaging Spectroradiometer (MISR) is a satellite instrument also designed to measure scattered sunlight upwelling from Earth and scheduled for launch into a polar orbit aboard NASA's EOS-AM. The MISR instrument uses nine individual CCD-based push-broom cameras to view Earth at nine different view angles: one at nadir and eight symmetrical views at 26.1, 45.6, 60.0, and 70.5 degrees forward and aft of nadir. Each camera will obtain images at four spectral bands centered at 443, 555, 670, and 865 nm with a horizontal resolution of 275 m, 550 m, or 1.1 km. The MISR data will be used to produce aerosol optical depth (Diner et al., 1998).

EOSP

The Earth Observing Scanning Polarimeter (EOSP) instrument proposes to measure radiance and linear polarization of reflected sunlight in 12 spectral bands from 0.41 to 2.25 µm. EOSP data will provide information on the global aerosol and cloud distribution and on such properties as optical depth, phase, particle size, and cloud-top pressure. At this time, EOSP is not funded for development and flight.

PICASSO-CENA

The Pathfinder Instruments for Clouds and Aerosols using Spaceborne Observations-Climatologie Etendue des Nuages et des Aerosols (PICASSO-CENA) instruments include a dual-wavelength (530 nm and 1060 nm), polarization-sensitive lidar, an oxygen A-band spectrometer operating over the O_2 absorption region at 763 to 769 nm with 0.5 cm^{-1} resolution and an imaging infrared radiometer operating at two wavelengths, 10.5 and 12 µm. In addition, a high-resolution, widefield camera will be boresighted with the lidar and the other instruments. PICASSO-CENA will fly in formation with the EOS-PM spacecraft, providing joint measurements within 6 minutes of each other. It is designed to address the overdependence of our present understanding of the climate system on theoretical models by providing data that will reduce uncertainties in aerosol and cloud forcing.

SCHIAMACHY

One non-U.S. instrument that will measure aerosols is the Scanning Imaging Absorption Spectrometer for Atmospheric Cartography (SCIAMACHY), to be launched by the European Space Agency (ESA) on the ENVISAT-1 in June 2001. The SCIAMACHY instrument is a spectrometer designed to measure sunlight transmitted, reflected, and scattered by Earth's atmosphere or surface in the ultraviolet, visible, and near infrared wavelength region (0.240 to 2.38 µm) at moderate spectral resolution. It is expected to provide aerosol optical depths and profiles of extinction and scattering with accuracies not yet determined.

Measurement Requirements and Capabilities

Uncertainties in aerosol radiative forcing must be quantified and reduced, or at least limited to ±1.5 Wm^{-2} if models are to provide accurate predictions of regional climate change. There is a consensus that anthropogenic sulfate aerosols produce a direct forcing that is a substantial fraction of the forcing from GHGs (NRC, 1996). At least three key parameters must be better quantified if the uncertainties in direct aerosol forcing are to be reduced: (1) optical depth (a measure of total column content), (2) single scattering albedo (the fraction of attenuated radiation that is scattered rather than absorbed), and (3) aerosol source strength (an essential input for models). Anthropogenic aerosol sources are located over land, and their forcings there are the largest. Since current spacecraft instruments cannot make tropospheric aerosol measurements over land, on a global scale, the uncertainties are large. Table 7.2 lists the expected optical depth measurement uncertainties for MODIS, MISR, and a number of other spaceborne sensors. Figure 7.2 (NRC, 1996) gives the optical depth accuracies needed for different optical depth values to reduce uncertainties in aerosol forcing to ±15 percent. Comparison of Table 7.2 and Figure 7.2 shows that only PICASSO-CENA is capable of meeting this optical depth requirement.

TABLE 7.2 Optical Depth Measurements and Radiative Forcing Uncertainty from Current and Future Satellite Instruments

Instrument[a]	Optical Depth Resolution/Uncertainty	Forcing[b] Resolution/Uncertainty (Wm^{-2})
GOME/ERS-2	0.05	1.5
SeaWiFS	0.03 over ocean	0.9
AVHRR/NOAA-K	0.05 over ocean	1.5
MODIS/TERRA	0.03 over ocean	0.9
MISR/TERRA	0.05 over ocean	1.5
EOSP/TERRA2	0.03	0.9
PICASSO-CENA	0.005	0.15
Global-mean, annually averaged aerosol forcing		0.4

[a]Acronyms for instruments are defined in Appendix B.
[b]Forcing estimates based on $\Delta f_r = 30 \, \Delta\tau$ Wm^{-2} (Harshvardhan, 1993).

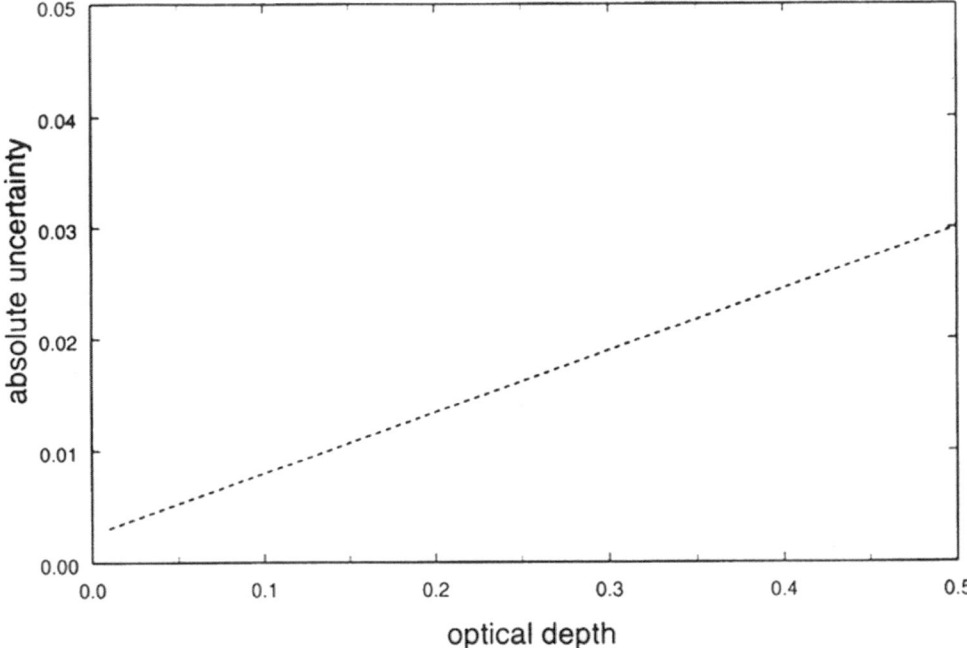

FIGURE 7.2 National Research Council (1996) estimates of optical depth accuracies versus optical depth values needed to reduce uncertainties in aerosol forcing to ±15 percent. Reprinted from NRC (1996).

Single-scattering albedo cannot currently be measured from space, and even in situ measurements are sparse and uncertain. New techniques or combinations of measurement techniques, both spaceborne and in situ, are needed to narrow significantly the uncertainty in the resulting aerosol forcing. Source strengths are also not well-known globally. They can be estimated from optical depth measurements but will remain a challenge for the currently approved nonlidar spaceborne experiments. Similarly, it will be difficult for spaceborne instruments to quantify indirect aerosol forcing. Again, a combination of instruments like PICASSO-CENA and EOS PM flying in formation, supported by carefully designed in situ measurements and aerosol modeling, will probably be able to reduce the large uncertainties in indirect aerosol forcing.

The National Polar-orbiting Operational Environmental Satellite System (NPOESS) program has proposed using the angstrom turbidity coefficient α as a representative aerosol size parameter (Table 7.3). The size parameter is valid only if the aerosol particle size distribution is given by an inverse power law, such as a Junge distribution. This is a reasonable assumption for volcanically unperturbed stratospheric conditions and very clean tropospheric conditions. During episodic periods of increased volcanic aerosol loading, the stratospheric size distribution becomes multimodal, and these size distributions will produce complex and nonlinear spectral shapes (Russell et al., 1996). The same is true for tropospheric aerosols near industrial regions. Thus, the size parameter α will not be useful for a volcanically perturbed stratosphere or a regionally polluted (industrial) troposphere, both of which must be taken into account in order to understand aerosol climate forcing.

The currently planned and funded satellite instruments described above have the potential to yield information on tropospheric and stratospheric aerosols. Without lidar or some other new approach, these systems are clearly limited in their application to understanding climate forcing. The techniques employed by the nadir-viewing passive systems are expected to work reliably over the oceans but to experience significant degradation over land and to provide no data on stratospheric aerosols in the presence of clouds. In addition, retrieval schemes for passive nadir instruments almost always require extensive a priori modeling of the aerosol's surface properties and of the aerosols themselves. Finally, because optical depths of the order of 0.05 must be measured to an accuracy of 10 to 20 percent, only SAGE III (~5 percent) in the stratosphere and spaceborne lidar in the troposphere, with supporting in situ measurements, can meet the accuracy demands at low and moderate optical depths. Nonetheless, the current and future spaceborne instruments, such as MODIS and MISR, are expected to enhance overall understanding of tropospheric aerosol characteristics globally.

CALIBRATION AND VALIDATION STRATEGY

The integrity of all satellite-derived data is based on calibrating and validating the sensors. Calibration is the process of quantitatively defining the system responses to known, controlled signal inputs and is critical for climate research. Prelaunch calibrations are required to establish a baseline for sensor operation. Postlaunch or

TABLE 7.3 Selected Environmental Data Record Requirements for the Measurement of Aerosol Size Parameters (α)

System Capability	Threshold	Objective
Vertical coverage	Surface to 30 km	Surface to 50 km
Horizontal coverage	Over ocean only	Global
Vertical resolution	Total column	
From 0 to 2 km		0.25 km
From 2 to 5 km		0.5 km
>5 km		1 km
Mapping uncertainty	4 km	1 km
Measurement range	−1 to +3	−2 to +4
Measurement precision	0.3	0.1
Measurement accuracy	±0.3 over ocean	±0.1
Refresh	6 hours	4 hours; 2 hours during daylight
Long-term stability (%)	0.1	0.03

SOURCE: Extracted from NOAA (1997) and IORD-1. IORD-1 and other documentation related to the NPOESS program are available online at <http://npoesslib.ipo.noaa.gov/ElectLib.htm>.

onboard calibrations are also a necessary component of sensor characterization. Validation involves evaluating the algorithms required to extract physical quantities from calibrated and well-characterized instrument products and is a means of quantifying the overall system calibration, including all the data-processing algorithms. Validation and calibration campaigns are crucial to analyzing potential inherent and systematic biases of newly launched and existing satellite systems. These campaigns are also important for algorithm development and quality control to ascertain potential biases and confidence in the retrieved quantities. Validation measurements must be maintained at some level of activity throughout the lifetime of the satellite experiment.

In the committee's view, a data validation plan should be organized to assess anticipated systematic uncertainties in the NPOESS aerosol data products and to better understand the assumptions underlying the measurement technique. The plan should be patterned after other successful validation programs. It should be based on intercomparisons with correlative measurements by in situ and remote sensors on the ground and aboard aircraft, balloons, and spacecraft (including EOS platforms). Intercomparisons are made with two kinds of data: planned measurements by in situ and remote sensors on a single-event basis and data from other sensors on a statistical or target-of-opportunity basis. Uncertainties in funding and in the long-term availability of aircraft and other forms of support require that the validation plan be a working document, periodically updated to reflect necessary changes.

Because there is no accepted technique for calibrating atmospheric aerosol extinction measurements, previous validation programs (e.g., SAM II and SAGE I and II) have relied on intercomparisons of satellite measurements with observations from lidars, Sun photometers, dustsondes, and particle samplers. Aerosol extinction is a derived product for these sensors, often requiring additional information such as a backscatter-to-extinction conversion factor or a knowledge of the aerosol composition or refractive index. Unfortunately, these derivations can introduce greater uncertainty into the intercomparison, so it is desirable to acquire an array of correlative measurements to gain closure on the retrieved products. For example, a more direct intercomparison will be possible between the inferred aerosol surface area observations and measurements of the size distribution from particle samplers.

Another important component of a validation plan is intercomparison of complementary observations by other measurement programs. This activity will allow ongoing intercomparisons to occur immediately following launch and to continue for the lifetime of the mission at little cost. It will further develop an important cross-reference data set for assessing biases between in situ and remote sensing instruments around the world. Satellite observations will also provide an important database for assessing instrument biases and precision. A commitment to long-term validation must be made so that global change can be quantified.

**Box 7.1
Findings**

In order to fulfill the need for a global data set of aerosol measurements, the committee believes that a limb occultation instrument operating in visible and near-infrared wavelengths, such as the Satellite Aerosol and Gas Experiment (SAGE) III in a highly inclined orbit, will provide the required stratospheric aerosol data record both in the transition period and in the National Polar-orbiting Operational Environmental Satellite System era. In addition, in the committee's estimation, spaceborne lidar complemented by other passive sensors offers the greatest likelihood of producing the data set required to understand the impact of tropospheric (and anthropogenic) aerosols on climate. The combination of a carefully planned in situ measurement network, a coordinated aircraft campaign, and simultaneous measurements from a synchronized satellite orbit using, e.g., the Moderate-Resolution Imaging Spectroradiometer (MODIS), and instruments boresighted with a lidar (e.g., an oxygen A-band instrument) on the same platform appears to be the ideal approach for depicting global tropospheric aerosols and reducing uncertainties in aerosol forcing to the required levels.

DATA MANAGEMENT

Reprocessing Issues

Usually, it becomes necessary to reprocess the entire data set and associated time series as the data record length increases. This ensures the integrity, consistency, and continuity of a high-quality data set suitable for climate research. As the satellite-based instruments age and (possibly) change their characteristics, as spacecraft orbits change, and as episodic events such as volcanic eruptions occur, the retrieval algorithms become outdated and it becomes evident that there is a need to continually compare measurements with those from newer instruments and updated retrieval algorithms. Funding must be made available for these activities. Raw data must also be stored so that new inversion techniques can later be tried or studied by inversion specialists. These needs become even more important for climate studies. To be usable by the science community, data (that is, derived products and satellite ephemeris data) must also be archived using readily accessible formats.

Algorithm Status

Algorithms used for studies of long-term change must be stable and capable of validation. Many satellite data sets are based either on empirical relationships (statistically determined from previous climatology) or on a priori assumptions, rather than on physically robust models. While the empirical approach is valuable, it may become obsolete after an episodic event such as a volcanic eruption. For example, Thomason (1991) used empirical relationships between the 0.525, 0.940, and 1.02 µm extinction measurements to seek a size distribution parameterization that gave the best water vapor retrieval at 0.940 µm. He showed that the SAGE II measurements of the pre-Pinatubo stratospheric aerosol could be modeled with size distributions in the form of a segmented power law while preserving the wavelength dependence of the measured extinction. The main problem with using a priori assumptions, such as a specific form of an aerosol size distribution, is that they may preclude other forms that may be more realistic or suitable.

Required Ancillary Data

The NPOESS instruments must provide meteorological or other data required for interpretation, validation, four-dimensional assimilation, and trajectory analyses. At a minimum, meteorological data such as temperature, pressure, and potential vorticity should be included as part of the data set. These three variables, for example, are useful for comparing satellite sensors with different vertical coordinates and for studying the transport mechanisms of the derived products on quasi-conserved surfaces.

EVOLUTION STRATEGY

There should be a continuing effort to develop and incorporate new technologies for the space-based observation of aerosols. Areas for research and development should include the following:

- New instrument techniques,
- Miniaturization,
- Use of lightweight components,
- Onboard data handling,
- Flexibility in instrument operations in space (for those that can be reconfigured),
- More capable detectors, noise-free and with higher quantum efficiency,
- More efficient and long-lived lasers or other such devices, and
- New output wavelengths.

Progress is also needed in areas such as the following that support the measurements needed to resolve cloud-aerosol forcing issues:

- Research in new remote sensors and in combinations of instruments, such as lidar and radar, for vertical profiling,
- Formation flying, and
- Group flying.

In general, better tropospheric measurements are also needed.

BIBLIOGRAPHY

Bekki, S. 1995. On the possible role of aircraft-generated soot in middle latitude ozone depletion. J. Geophys. Res. 100: 7195.

Bevilacqua, R.M. 1997. Introduction to special section: Polar Ozone and Aerosol Measurement (POAM II). J. Geophys. Res. 102: 23591-23592.

Blake, D.F., and K. Kato. 1995. Latitudinal distribution of black carbon soot in the upper troposphere and lower stratosphere. J. Geophys. Res. 100: 7202.

Charlson, R.J., J. Langner, H. Rodhe, C.B. Loevy, and S.G. Warren. 1991. Perturbation of the northern hemisphere radiative balance by backscattering from anthropogenic sulfate aerosols. Tellus 43AB: 152-163.

Charlson, R.J., S.E. Schwartz, J.M. Hales, R.D. Cess, J.A. Coakley Jr., J.E. Hansen, and D.J. Hoffman. 1992. Climate forcing by anthropogenic aerosols. Science 255: 423-430.

Chin, M., and D.D. Davis. 1993. Global sources and sinks of OCS and CS_2 and their distributions. Global Biogeochemical Cycles 7: 321-337.

Chin, M., and D.D. Davis. 1995. A reanalysis of carbonyl sulfide as a source of stratospheric background sulfur aerosol. J. Geophys. Res. 100: 8993-9005.

Chu, W.P., M.P. McCormick, J. Lenoble, C. Brogniez, and P. Pruvost. 1989. SAGE II inversion algorithm. J. Geophys. Res. 94: 8339-8351.

Crutzen, P.J. 1976. The possible importance of CSO for the sulfate layer of the stratosphere. Geophys. Res. Lett. 3: 73-76.

Debrestian, D.J., J.D. Lumpe, E.P. Shettle, R.M. Bevilacqua, J.J. Olivero, J.S. Hornstein, W. Glaccum, D.W. Rusch, C.E. Randall, and M.D. Fromm. 1997. An analysis of POAM II solar occultation observations of polar mesospheric clouds in the southern hemisphere. J. Geophys. Res. 102: 1971-1981.

Diner, D.J., J.C. Beckert, T.H. Reilly, C.J. Bruegge, J.E. Conel, R. Kahn, J.V. Martonchik, T.P. Ackerman, R. Davies, S.A.W. Gerstl, H.R. Gordon, J-P. Muller, R. Myneni, R.J. Sellers, B. Pinty, and M.M. Verstraete. 1998. Multi-angle Imaging Spectroradiometer (MISR) instrument description and experiment overview. IEEE Trans. Geosci. Remote Sensing 36: 1072-1085.

Drdla, K. 1996. Applications of a model of polar stratospheric clouds and heterogeneous chemistry. Ph.D. thesis, University of California at Los Angeles.

Duce, R.A. 1995. Sources, distributions, and fluxes of mineral aerosols and their relationship to climate. Aerosol Forcing of Climate, R.J. Charlson (ed.). New York: John Wiley & Sons.

Durkee, P.A., et al. 1991. Global analysis of aerosol particle characteristics. Atmos. Environ. 25A: 2457-2471.

Esaias, W., M.R. Abbott, I. Barton, O.B. Brown, J.W. Campbell, K.L. Carder, D.K. Clark, R.H. Evans, F.E. Hoge, H.R. Gordon, W.M. Balch, R. Letelier, and P.J. Minnett. 1998. An overview of MODIS capabilities for ocean science observations. IEEE Trans. Geosci. Remote Sensing 36: 1250-1264.

Fromm, M.D., R.M. Bevilacqua, J.D. Lumpe, E.P. Shettle, J.S. Hornstein, S.T. Massie, and K.H. Fricke. 1997. Observations of Antarctic polar stratospheric clouds by POAM II: 1994-1996. J. Geophys. Res. 102: 23659-23672.

Glaccum, W., R. Lucke, R.M. Bevilacqua, E.P. Shettle, J.S. Hornstein, D.T. Chen, J.D. Lumpe, S.S. Krigman, D.J. Debrestian, M.D. Fromm, F. Dalaudier, E. Chassefiere, C. Deniel, C.E. Randall, D.W. Rusch, J.J. Olivero, C. Brogniez, J. Lenoble, and R. Kremer. 1996. The Polar Ozone and Aerosol Measurement instrument. J. Geophys. Res. 101: 14479-14487.

Griggs, M. 1975. Measurement of atmospheric aerosol optical thickness over water using ERTS-1 Data. J. Air Pollut. Control Assoc. 25: 622-626.

Hansen, J.E., M. Sato, A. Lacis, R. Ruedy, I. Tengen, and E. Matthews. 1998. Climate forcings in the industrial era. Proc. Natl. Acad. Sci. U.S.A. 95: 12753-12758.

Harshvardhan. 1993. Aerosol-Climate Interactions. Aerosol-Cloud-Climate Interactions. P.V. Hobbs (ed.). San Diego: Academic Press, p. 81.

Haywood, J.M., and K.P. Shine. 1995. The effect of anthropogenic sulfate and soot aerosol on the clear sky planetary radiation budget. Geophys. Res. Lett. 22: 603-606.

Haywood, J.M., V. Ramaswamy, and B.J. Soden. 1999. Tropospheric aerosol climate forcing in clear-sky satellite observations over the oceans. Science 283:1299-1303.

Herman, M., J.L. Deuzé, C. Devaux, Ph. Goloub, F.M. Bréon, and D. Tanré. 1997. Remote sensing of aerosols over land surfaces, including polarization measurements; application to some airborne POLDER measurements. J. Geophys. Res. 102: 17039-17049.

Hervig, M.E., T. Deshler, and J.M. Russell III. 1998. Aerosol size distributions obtained from HALOE spectral extinction measurements. J. Geophys. Res. 103: 1573-1583.

Hofmann, D.J. 1990. Increase in the stratospheric background sulfuric acid aerosol mass in the past 10 years. Science 248: 996-1000.

Holben, B.N., V. Kalb, Y.J. Kaufman, D. Tanré, and E. Vermote. 1992. Aerosol retrieval over land from AVHRR data—Application for atmospheric correction. IEEE Trans. Geosci. Remote Sensing 30: 212-222.

Holben, B.N., T.F. Eck, I. Slutsker, D. Tanre, J.P. Buis, A. Setzer, E. Vermote, J.A. Reagan, Y.J. Kaufman, T. Nakajima, F. Lavenu, I. Jankowiak, and A. Smirnov. 1998. AERONET—A federated instrument network and data archive for aerosol characterization. Remote Sens. Environ. 66: 1-16.

Ignatov, A.M., L.L. Stowe, S.M. Sakerin, and G.K. Korotaev. 1995. Validation of the NOAA/NESDIS satellite aerosol products over the North Atlantic in 1989. J. Geophys. Res. 100: 5123-5132.

Intergovernmental Panel on Climate Change (IPCC). 1995. Climate Change 1995: The Science of Climate Change, J.T. Houghton, L.G. Meira Filho, B.A. Callendar, N. Harris, A. Kattenberg, and K. Maskell (eds.), New York: Cambridge University Press.

International Global Atmospheric Chemistry Project (IGAC). 1996. Atmospheric aerosols: a new focus of the International Global Atmospheric Chemistry Project, P.V. Hobbs and B.J. Huebert (eds.), IGAC Core Project Office, Mass. Inst. Technol., Cambridge.

Junge, C.E., C.W. Chagnon, and J.E. Manson. 1961. Stratopheric aerosols. J. Meteorol. 18: 81-108.

Karl, G.S., et al. 1995. Evidence for radiative effects of anthropogenic sulfate aerosols in the observed climate record. In Aerosol Forcing of Climate, R.J. Charlson and J. Heintzenberg (eds.). New York: John Wiley & Sons.

Kaufman, Y.J. 1995. Remote sensing of direct and indirect aerosol forcing, pp. 297-332 in Aerosol Forcing of Climate. R.J. Charlson and J. Heintzenberg (eds.). Chichester, U.K.: John Wiley & Sons.

Kent, G.S., U.O. Farrukh, P.H. Wang, and A. Deepak. 1988. SAGE I and SAM II measurements of 1-m aerosol extinction in the free troposphere. J. Appl. Meteorol. 27: 269-279.

Khalil, M.A.K., and R.A. Rasmussen. 1984. Global sources, lifetimes and mass balances of carbonyl sulfide (OCS) and carbon-disulfide (CS_2) in the Earth's atmosphere. Atmos. Environ. 18: 1805-1813.

Kiehl, J.T., and B.P. Briegleb. 1993. The relative roles of sulfate aerosols and greenhouse gases in climate forcing. Science 260: 311-314.

Mauldin, L.E. III, N.H. Zaun, M.P. McCormick, J.H. Guy, and W.R. Vaughn. 1985. Stratospheric Aerosol and Gas Experiment II instrument: A functional description. Opt. Eng. 24: 307-312.

McCormick, M.P., H.M. Steele, P. Hamill, W.P. Chu, and T.J. Swissler. 1982. Polar stratospheric cloud sightings by SAM II. J. Atmos. Sci. 39: 1387-1397.

McCormick, M.P. 1987. SAGE II: An overview. Adv. Space Res. 7: 219-226.

McCormick, M.P., D.M. Winker, E.V. Browell, J.A. Coakley, C.S. Gardner, R.M. Hoff, G.S. Kent, S.H. Melfi, R.T. Menzies, C.M.R. Platt, D.A. Randall, and J.A. Reagan. 1993. Scientific investigations planned for the lidar in-space technology experiment (LITE). Bull. Am. Meteorol. Soc. 74: 205-214.

McCormick, M.P., L.W. Thomason, and C.R. Trepte. 1995. Atmospheric effects of the Mt. Pinatubo eruption. Nature 373: 393-404.

McCormick, M.P., W.P. Chu, and L.E. Mauldin. 1999. The flight of SAGE III on ISS, pp. 205-210 in Space Technology and Applications International Forum. M.S. El-Genk (ed.).

McElroy, M.B., R.J. Salawitch, S.C. Wofsy, and J.A. Logan. 1986. Reduction of Antarctic ozone due to synergistic interactions of chlorine and bromine. Nature 321: 759-762.

National Oceanic and Atmospheric Administration (NOAA). 1997. Climate Measurement Requirements for the National Polar-orbiting Operational Environmental Satellite System (NPOESS): Workshop Report, Herbert Jacobowitz (ed.), Office of Research Applications, NESDIS-NOAA, February. 77 pp.

National Research Council (NRC). 1996. Aerosol Radiative Forcing and Climate Change. Washington, D.C.: National Academy Press.

Novakov, T., and J.E. Penner. 1993. Large contributions of organic aerosols to cloud condensation nuclei concentrations. Nature 365: 823-826.

Penner, J.E., R.J. Charlson, J.M. Hales, N.S. Laulainen, R. Leifer, T. Novakov, J. Ogren, L. F. Radke, S. E. Schwartz, and L. Travis. 1994. Quantifying and minimizing uncertainty of climate forcing by anthropogenic aerosols. Bull. Am. Meteorol. Soc. 75(3): 375-400.

Pueschel, R.F. 1996. Stratospheric aerosols—formation, properties, effects. J. Aerosol Sci. 27: 383-402.

Pyle, D.M., P.D. Beattie, and G.J.S. Bluth. 1996. Sulfur emissions to the stratosphere from explosive volcanic eruptions. Bull. Volcanol. 57: 663-671.

Randall, C.E., D.W. Rusch, J.J. Olivero, R.M. Bevilacqua, L.R. Poole, J.D. Lumpe, M.D. Fromm, K.W. Hoppel, J.S. Hornstein, and E.P. Shettle. 1996. An overview of POAM II aerosol measurements at 1.06 m. Geophys. Res. Lett. 23: 3195-3198.

Rao, C.R.N., L.L. Stowe, and E.P. McClain. 1989. Remote sensing of aerosols over the oceans using AVHRR data: Theory, practice, and applications. Int. J. Remote Sensing 10: 743-749.

Rinsland, C.P., M.R. Gunson, R.J. Salawitch, H.A. Michelsen, R. Zander, M.J. Newchurch, M.M. Abbas, M.C. Abrams, G.L. Manney, A.Y. Chang, F.W. Irion, A. Goldman, and E. Mahieu. 1996. Trends of OCS, HCN, SF_6, $CHCLF_2$ (HCFC-22) in the lower stratosphere from 1985 and 1994 atmospheric trace molecule spectroscopy experiment measurements near 30 degrees N latitude. Geophys. Res. Lett. 23: 2349-2352.

Russell, J.M. III, L.L. Gordley, J.H. Park, S.R. Drayson, D.H. Hesketh, R.J. Cicerone, A.F. Tuck, J.E. Frederick, J.E. Harries, and P.J. Crutzen. 1993. The halogen occultation experiment. J. Geophys. Res. 98: 10777-10797.

Russell, P.B., J.M. Livingston, R.F. Pueschel, J.J. Hughes, J.B. Pollack, S.L. Brooks, P.J. Hamill, L.W. Thomason, L.L. Stowe, T. Deshler, E.G. Dutton, and R.W. Bergstrom. 1996. Global to microscale evolution of the Pinatubo aerosol, derived from diverse measurements and analyses. J. Geophys. Res. 101: 18745-18763.

Russell, P.B., P.V. Hobbs, and L.L. Stowe. 1999. Aerosol properties and radiative effects on atmospheric radiation in the United States East Coast haze plume: an overview of the Tropospheric Aerosol Radiative Forcing Observational Experiment (TARFOX). J. Geophys. Res. 104: 2213-2222

Solomon, S., R.R. Garcia, F.S. Rowland, and D.J. Wuebbles. 1986. On the depletion of Antarctic ozone. Nature 321: 755-758.

Solomon, S., R.W. Portman, R.R. Garcia, L.W. Thomason, L.R. Poole, and M.P. McCormick. 1996. The role of aerosol variations in anthropogenic ozone depletion at northern midlatitudes. J. Geophys. Res. 101: 6713-6727.

Stothers, R.B. 1996. Major optical depth perturbations to the stratosphere from volcanic eruptions: pyrheliometric period, 1881-1960. J. Geophys. Res. 101: 3920.

Stowe, L.L., R.M. Carey, and P.P. Pellegrino. 1992. Monitoring the Mt. Pinatubo aerosol layer with NOAA/11 AVHRR data. Geophys. Res. Lett. 19: 159-162.

Stratospheric Processes and their Role in Climate (SPARC). 1998. Newsletter number 10. Published by Meteo-France, January.

Tanré, D., Y.J. Kaufman, M. Herman, and S. Mattoo. 1997. Remote sensing of aerosol properties over oceans using the MODIS/EOS spectral radiances. J. Geophys. Res. 102: 16971-16988.

Thomason, L.W. 1991. A diagnostic stratospheric aerosol size distribution inferred from SAGE II measurements. J. Geophys. Res. 96: 22501-22508.

Thomason, L.W., and L.R. Poole. 1993. Use of stratospheric aerosol properties as diagnostics of Antarctic vortex processes. J. Geophys. Res. 98: 23003-23012.

Torres, O., P.K. Bhartia, J.R. Herman, Z. Ahmad, and J. Gleason. 1998. Derivation of aerosol properties from satellite measurements of backscattered ultraviolet radiation: theoretical basis. J. Geophys. Res. 103: 17099.

Weisenstein, D.K., G.K. Yue, M.K.W. Ko, N.-D. Sze, J.M. Rodriguez, and C.J. Scott. 1997. A two-dimensional model of sulfur species and aerosols. J. Geophys. Res. 102: 13019-13035.

Zander, R., C.P Rinsland, C.B. Farmer, J. Namkung, R.H. Norton, and J.M. Russell III. 1988. Concentration of carbonyl sulfide and hydrogen cyanide in the free troposphere and lower stratosphere deduced from ATMOS/Spacelab 3 infrared solar occultation spectra. J. Geophys. Res. 93: 1669-1678.

8

Ozone

INTRODUCTION

Atmospheric ozone has several environmental implications, which can be classified roughly by altitude:

- *In the stratosphere*, where 90 percent of atmospheric ozone resides, ozone plays a critical role in absorbing ultraviolet (UV) radiation and preventing it from reaching Earth's surface.
- *In the upper and middle troposphere*, ozone is a major greenhouse gas, causing inhomogeneous radiative forcing.
- *In the lower and middle troposphere*, ozone maintains the oxidizing power of the atmosphere by providing a source of the hydroxyl radical (OH^-) in the presence of water vapor. Oxidation by OH^- is the main sink for a number of environmentally important gases, including methane (CH_4), carbon monoxide (CO), hydrofluorocarbons, and methyl bromide.
- *In surface air*, ozone is a pernicious pollutant, toxic to humans and vegetation. It is the principal contributor to smog over the United States.

Anthropogenic emissions affect ozone in a complicated way involving nonlinear chemical processes and transport over a wide range of scales. It is well established that human activity has caused a decrease of ozone in the stratosphere and an increase in the troposphere, but the mechanisms are still unclear and predictions for the future are uncertain. The importance of continuously monitoring ozone trends throughout the atmosphere has long been recognized, and regular reports are published by the World Meteorological Organization (WMO, 1999). The role of space-based measurements as a key component of trend assessments is well established. The National Polar-orbiting Operational Environmental Satellite System (NPOESS) includes as one of its environmental data record (EDR) requirements the measurement of columns and vertical profiles of ozone with the Ozone Mapping and Profiler Suite (OMPS). Important EDRs for OMPS are summarized in Table 8.1.

The committee's findings concerning the current status and future NPOESS plans for measurement of ozone in the stratosphere and troposphere for research purposes are given in Box 8.1.

TABLE 8.1 Selected Environmental Data Record Requirements for the NPOESS Ozone Mapping and Profiler Suite

System Capability	Threshold	Objective
Horizontal resolution		
Total column	50 km at nadir	50 km worst case
Vertical profile	250 km	250 km
Vertical resolution		
0-10 km	N/A	3 km
10-25 km	5 km	1 km
25-60 km	5 km	3 km
Measurement precision		
Total column	0.001 atm-cm	0.001 atm-cm
Profile		
0-10 km	N/A	10%
10-15 km	10%	3%
15-50 km	3%	1%
50-60 km	10%	3%
Measurement accuracy		
Total column	±0.015 atm-cm	±0.005 atm-cm
Profile		
0-10 km	N/A	10%
10-15 km	20%	10%
15-60 km	10%	5%
Long-term stability (%)		
Total column	1%	0.5%
Profile	2%	1%

SOURCE: IPO NPOESS (1996). The updated IORD and other documentation related to the NPOESS program are available online at <http://npoesslib.ipo.noxaa.gov/ElectLib.htm>.

BASIC SCIENCE ISSUES

Long-Term Trend in Ozone: The Measurement Record

The environmental implications of stratospheric ozone depletion were just emerging when the Nimbus-7 satellite was launched in 1978. This marked the beginning of space-based ozone monitoring with observations from the Solar Backscatter Ultraviolet (SBUV), Total Ozone Mapping Spectrometer (TOMS), and Satellite Aerosol and Gas Experiment (SAGE) instruments. Since that time a National Plan for Stratospheric Monitoring (NOAA, 1989) was put into place. Its keystone was the continuation of ozone observations on the NOAA polar orbiting satellite series with the SBUV/2 instrument. Parallel to this plan, NASA has been flying improved TOMS and SAGE instruments on U.S. and Russian environmental satellites. The Upper Atmosphere Research Satellite (UARS) launched in 1991 carried instruments to measure several stratospheric parameters, including ozone. The Global Ozone Monitoring Experiment (GOME), a new multispectral nadir instrument launched on the European Earth Resources Satellite (ERS-2) in April 1995, is expected to provide data on tropospheric ozone columns and related constituents.

These space-based observations are supplemented by several other long-term data sets. The Dobson and Brewer networks of ground-based spectrophotometers provide total ozone column measurements at more than 150 sites. Ozonesondes launched from sites around the world, at frequencies that vary from twice a month to three times a week, provide detailed vertical profiles of ozone from the surface to the middle stratosphere (Logan, 1999; Logan et al., 1999). Aircraft campaigns sponsored by NASA and other agencies measure ozone concurrently with

> **Box 8.1**
> **Findings**
>
> Current understanding of atmospheric ozone has been reviewed by WMO (1999). Three priorities for future space-based observations emerge from the WMO report: (1) improved observation of ozone concentrations at altitudes below 20 km, (2) measurement of ozone concurrently with related species, and (3) better understanding of the long-term trend in ozone. How do NASA and NPOESS plans meet the science requirements?
>
> The research plans of NASA target the first two of these priorities but not the third; there is no commitment by the agency to ensure continuity of observations, as is needed for long-term trend studies. Indeed, the ability to detect long-term trends from satellite observations over the past decade has been largely serendipitous, and yet it has proven crucial for assessing human effects on atmospheric ozone.
>
> The NPOESS environmental data record (EDR) objectives for the operational Ozone Mapping and Profiler Suite (OMPS) instrument (see Table 8.1) provide specifications comparable to those of the new generation of research instruments and would make NPOESS a powerful source of information for detecting long-term trends in atmospheric ozone. The EDR thresholds (Table 8.1) are sufficient for detecting long-term trends in ozone columns but inadequate for detecting trends in the vertical distribution of ozone because of the coarse vertical resolution (5 km) and the insufficient precision below 15 km. Resolving the vertical distribution of ozone trends is critical to interpreting trends in the total column and assessing ozone radiative forcing.
>
> The following observations for the OMPS sensor on NPOESS are intended to ensure its usefulness for monitoring long-term trends in ozone:
>
> - The OMPS should significantly exceed the EDR thresholds and provide or approach the EDR objectives below 25 km.
> - There should be a 1-year overlap between successive OMPS launches to allow sensor intercomparison and guarantee long-term traceability.
> - Calibration and validation of the OMPS must be viewed as a critical activity to be maintained throughout the lifetime of the instrument. It should be led by a group independent of the OMPS team.
> - Any changes made to the retrieval algorithm should be followed by reprocessing of the entire record of OMPS observations to preserve the integrity of the record for long-term trend analyses. This requirement implies in particular that the raw radiances from OMPS should be archived.

a large number of related species to improve our understanding of the factors controlling ozone in the lower stratosphere and in the troposphere.

The recent WMO (1999) report gives a detailed discussion of the measurement capabilities of the space- and ground-based instruments used to assess long-term trends in ozone concentration. The principal instruments are summarized in Table 8.2. The accuracies, precisions, and instrument drifts in Table 8.2 were verified by intercomparisons with other ground- and space-based instruments (WMO, 1999). Additional space-based data for ozone are available from the TIROS-N operational vertical sounders (TOVS) on NOAA polar-orbiting satellites from 1979 to the present. Comparison with other sounders indicates that TOVS are sensitive only to trends in lower stratospheric ozone. TOVS ozone retrievals are complicated by cloud effects, water vapor absorption, and surface emissivity; intersatellite instrument differences further complicate the interpretation of the data. Because of these problems, TOVS data have been of little use for long-term trend analyses.

Trends in total ozone columns for the period 1978 through 1998 have been analyzed using data from the TOMS and SBUV instruments as well as from the Dobson and Brewer ground-based networks (WMO, 1999). The different records are in good agreement:

TABLE 8.2 Measurement Capabilities of Ozone Instruments Currently Used for Long-Term Trend Analyses

Instrument[a]	Measurement	Accuracy (%)	Precision (%)	Time-dependent drift error (%)
Dobson and Brewer Spectrophotometers	Total column	1-2	0.5 (annual mean)	
TOMS	Total column	3	2	1.5 (14 yr)
SBUV, SBUV/2	Vertical profile (25-45 km, 5-km resolution)	3	2	3
SAGE, SAGE-II	Vertical profile (20-50 km, 1-km resolution)	3-5	2	<0.5 yr^{-1}
Ozonesondes	Vertical profile (0-30 km, high resolution)	3-5	3	

[a]Acronyms for instruments are defined in Appendix B.

- No significant trends in tropical regions (20 degrees S to 20 degrees N);
- Negative trends in the extratropical northern hemisphere of 3 to 6 percent per decade in winter and spring and 1 to 3 percent per decade in summer and fall;
- Large negative trends of 6 to 22 percent per decade at high southern latitudes in winter and spring, due to the influence of the Antarctic ozone hole, with weaker negative trends of 2 to 5 percent per decade in summer; and
- Recent declines in Arctic springtime, with ozone loss during individual months of 25 to 35 percent in 1996 and 1997.

The origin of the Antarctic ozone hole is now well understood, but there is still much uncertainty about the trends in the midlatitudes and in the Arctic. Examining the vertical profile in the trend affords some insight. Space-based observations of the vertical distribution of ozone are available from the SAGE and SBUV instruments but do not extend reliably below 20 km altitude. Ozonesondes offer the only source of information at lower altitudes, but with poor spatial coverage, mainly over extratropical continental regions of the northern hemisphere. Data compiled in the WMO (1999) assessment indicate that ozone trends are negative at all altitudes from 10 to 65 km, with two local maxima: 7.4±2.0 percent per decade at 40 km and 7.6±4.6 percent per decade at 15 km (Figure 8.1). Most of the decline in the total ozone column takes place below 20 km, where ozonesonde data are the only data available. There is a need to extend satellite observations to that altitude range and also to improve the long-term calibration of the ozonesonde measurements.

Long-term trends in tropospheric ozone can be determined at only a few northern midlatitude stations where there is sufficient ozonesonde coverage. Data for 1970 through 1996 show decreases or no significant trends at Canadian stations, no significant trends at U.S. stations, and increases of 5 to 15 percent per decade at European and Japanese stations (WMO, 1999). For 1980 through 1996, there are generally no significant trends except for decreases at the Canadian stations of 2 to 8 percent per decade. Increasing subsonic aircraft emissions have so far not produced a detectable trend of ozone in the upper troposphere (Logan, 1994). According to atmospheric chemistry models, the current subsonic aircraft fleet should have caused a 3 to 9 percent increase in ozone at 9 to 13 km at northern midlatitudes, but such a change could have gone undetected in trend analyses because of other factors influencing ozone. A further increase of 13 percent in ozone in the upper troposphere at northern midlatitudes is expected by 2050 owing to continued increases in aircraft emissions (IPCC, 1999).

Where Is the Science Heading?

The main science questions relating to ozone trends over the next 20 years are expected to be the following:

- *Polar regions:* Will Antarctic ozone levels recover as chlorine levels decline? Will ozone levels in the Arctic continue to decline? How will changes in stratospheric climate affect polar ozone loss?

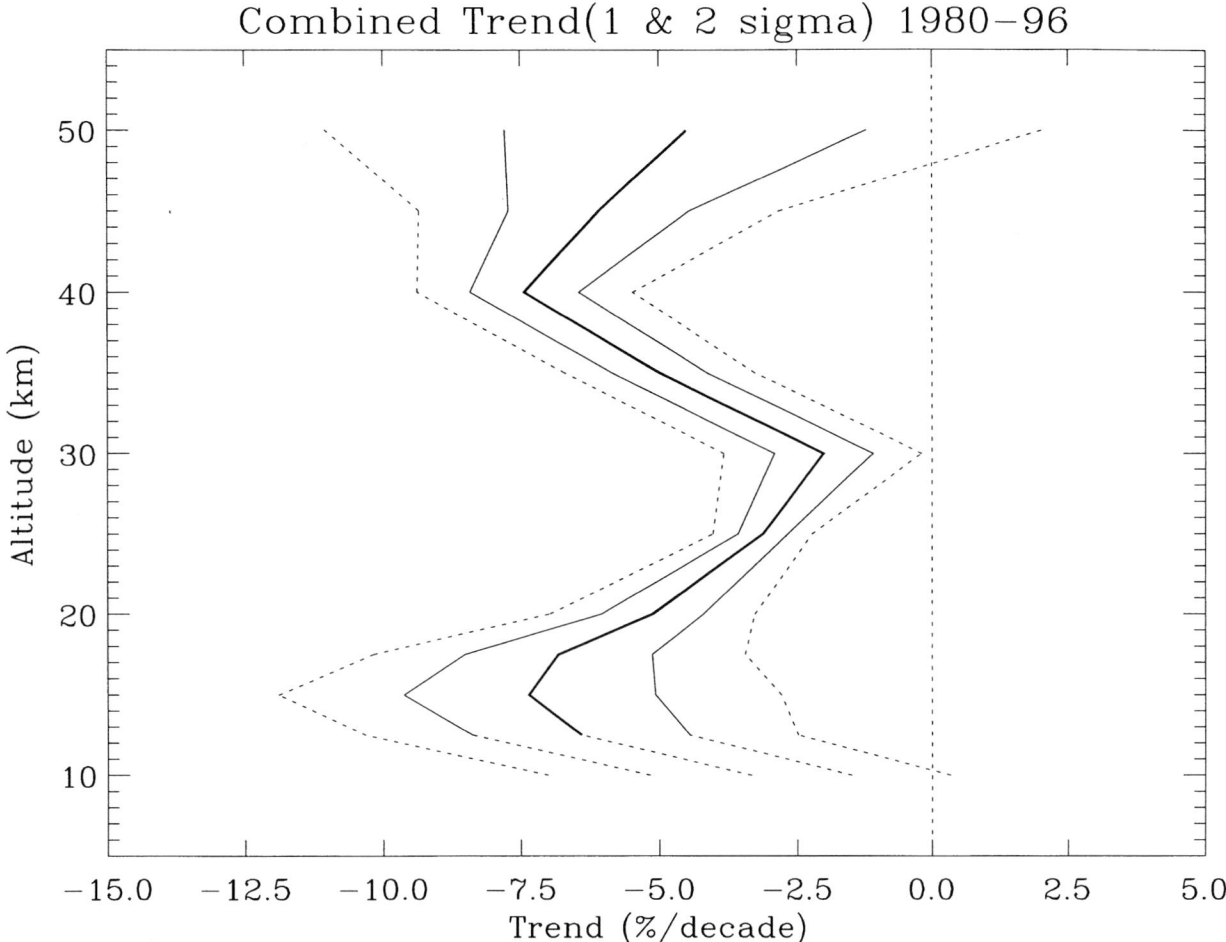

FIGURE 8.1 Estimate of mean ozone trend for 1980 to 1996 using data from Umkehr, ozonesondes, SAGE I/II, and SBUV measurement systems at the northern and midlatitudes (heavy solid line). Combined uncertainties are also shown as 1 sigma (light solid line) and 2 sigma (dashed line). Combined trends and uncertainties are extended down to 10 km as shown by the light dotted lines. The results below 15 km are a mixture of tropospheric and stratospheric trends, and the exact numbers should be viewed with caution. Combined trends have not been extended lower into the troposphere because the small sample of sonde stations introduces additional unqualified uncertainty about the representativeness of mean trends. (Reprinted from WMO, 1999.)

- *Northern midlatitudes:* Will ozone columns continue to decline or will they recover?
- *Troposphere:* How will ozone concentrations change in response to changes in subsonic aircraft emissions, industrial emissions at northern midlatitudes, and biomass burning and economic development in the tropics?

Measurement Requirements

Instrument Capabilities

The ability to discern future trends in ozone from space-based sensors will depend on the magnitude of the trend, the length of the observational record, and the degree of autocorrelation and noise in the data. A reasonable requirement is that a trend of 2 percent per decade be detectable from a 15-year record. To meet such a requirement, the noise in the data must be less than 3 percent (Weatherhead et al., 1998), which places a lower limit on the instrument precision. In addition, a 1-km vertical resolution is needed to resolve the fine vertical structure of the trend (see Figure 8.1). Resolving this structure is important for interpreting the trend for chlorine, aerosols, and other changes.

Overlapping Observations

Since 1978, there have been near-continuous space-based observations of ozone columns and stratospheric ozone profiles from a combination of TOMS, SBUV, and SAGE, supplemented more recently by measurements from the UARS satellite (the Halogen Occultation Experiment (HALOE) and the Microwave Limb Sounder (MLS) instruments) and from GOME. Temporal overlaps between these instruments have allowed detailed intercomparisons to play a key role in assessing the precision, accuracy, and long-term drift of the instruments (WMO, 1999). However, these overlaps have been somewhat serendipitous; no deliberate effort has been made to ensure the continuity and long-term traceability of ozone measurements.

Gaps in the satellite record are unacceptable for detecting long-term trends in ozone, because satellites offer the only means for global monitoring of the ozone distribution. Ground-based observations from spectrophotometers, sondes, or lidars offer only sparse spatial coverage. They are inadequate for observing, for example, the evolution of polar ozone loss in the Arctic, an issue that is emerging as a major environmental concern for the decades ahead. Continuity of space-based observations of ozone, involving planned overlap between successive instruments, is essential for detecting and interpreting long-term trends.

Much of the emphasis in long-term ozone monitoring has been on the total ozone column, which is directly and simply related to trends in the ultraviolet flux at the surface. As discussed previously, high-precision measurement of the vertical distribution is critical for interpreting trends in the total ozone column. There is currently some uncertainty in the ozone trend below 20 km because of inadequate measurement capabilities. This situation should improve with the new generation of space-based instruments.

There is a pressing need to develop a capability for space-based detection of trends in tropospheric ozone. Ozonesondes are currently the only means for diagnosing ozone trends in the troposphere, but the sparsity of the ozonesonde network lends considerable uncertainty to trend assessments. The new generation of satellite instruments will improve the capability for space-based observations of ozone in the troposphere and in the tropopause region. However, these instruments are intended for research applications and have short planned lifetimes. There is no plan at this time to use space-based measurements to detect long-term trends in tropospheric ozone, and this gap is reflected by the absence of an NPOESS threshold EDR for ozone below 10 km altitude (see Table 8.1). In the committee's view, however, technology for reliable space-based measurement of tropospheric ozone, at least down to 5 km altitude, should nevertheless be available for operational use by the time NPOESS is launched.

Data Management

Calculation of ozone columns and vertical profiles from the radiances measured aboard a satellite is based on well-established radiative transfer theory, and the algorithms used by the current generation of sensors can be described as mature. It is essential for long-term traceability of the data that any changes in the algorithms over the lifetime of an instrument be accompanied by a complete reprocessing of the previously collected data. This reprocessing is done routinely for the existing suite of ozone sensors (e.g., TOMS, SAGE II) and should be adhered to in managing NPOESS data. Such routine reprocessing requires that the raw radiance data from the sensors must be archived.

OBSERVING STRATEGY

NASA and NPOESS/IPO Plans

The next ozone sensor planned for launch by NASA will be the SAGE III instrument in a Sun-synchronous orbit, possibly aboard a Russian spacecraft. SAGE III will have better performance and dynamic range than SAGE II, especially below 20 km, and it will have additional capabilities for measuring water vapor, nitrogen dioxide (NO_2), aerosols, and subvisible cirrus (SAGE III is described in Chapter 7 of this report). The Sun-synchronous orbit will provide SAGE III measurements in both polar regions. Global coverage of the vertical distribution of ozone will require an additional SAGE III instrument flying in an inclined orbit. There are plans to fly a SAGE III instrument on the International Space Station in an inclined orbit (51.6 degrees) beginning in 2002. An earlier launch in an inclined orbit is being considered by NASA as a flight of opportunity, but a spacecraft has not been identified. The delay in achieving global coverage with SAGE III is a serious concern, as the existing Earth Radiation Budget Satellite (ERBS) spacecraft carrying the SAGE II experiment in an inclined orbit has been in space since 1984 and could cease operation any day. As previously mentioned, if SAGE II were lost, continuity in the data record for the vertical distribution of ozone would be lost, as well as the opportunity to attribute ozone trends in coming years to aerosol changes, chlorine, and other factors.

The Earth Observing System (EOS)-Chemistry satellite mission, scheduled for launch in December 2002 on a polar orbit, will have a payload focused on improving our understanding of ozone in the stratosphere and the troposphere. The payload will include:

• The Ozone Monitoring Instrument (OMI), an instrument of TOMS and GOME heritage, which will measure column concentrations of ozone and a number of other species.

• The High Resolution Dynamics Limb Sounder (HIRDLS), an infrared limb-scanning radiometer designed to sound the upper troposphere, stratosphere, and mesosphere for ozone, a number of ancillary gases, and aerosols. HIRDLS will provide sounding observations with horizontal and vertical resolution superior to that previously obtained and will observe the lower stratosphere with improved sensitivity and accuracy.

• The Microwave Limb Sounder, which will measure the concentrations of ozone and a number of ancillary gases in the stratosphere and will also measure ozone, water vapor, and cirrus ice content in the upper troposphere. An earlier version of MLS was flown on the UARS satellite.

• The Tropospheric Emission Spectrometer (TES), a Fourier transform infrared spectrometer that will measure the emission of ozone and other species, including water vapor, CO, nitric oxide, and nitric acid (HNO_3) at high spectral resolution in the infrared. TES will have both nadir and limb observation capabilities and will focus particularly on tropospheric observations, where it will provide vertical resolutions of 2.3 km (limb) and about 4 km (nadir).

NPOESS includes as one of its EDRs the measurement of columns and vertical profiles of ozone with the OMPS. One OMPS flight unit is to be provided on the Polar-orbiting Operational Environmental Satellite in 2004, to be followed by three flight units for NPOESS launches in 2007, 2010, and 2016. Comparison with future requirements for detecting long-term trends in ozone (discussed previously) indicates that an OMPS instrument meeting the EDR objective would provide an excellent record for assessing trends in ozone columns and profiles down to 10 km. An OMPS instrument that simply meets the threshold would be inadequate to assess trends below 25 km because of the coarse vertical resolution and low precision. As discussed above, vertical resolution of trends below 25 km is essential for interpreting trends in total ozone columns. There is, therefore, a considerable difference between the threshold and objective EDRs in terms of the usability of OMPS data for long-term trend monitoring. This usability will also be contingent on the overlap of OMPS records from successive NPOESS satellites to ensure long-term traceability. An overlap of at least 1 year is desirable to provide comparison data for a full annual cycle (R. McPeters, Goddard Space Flight Center, personal communication, 1998).

International Plans

Two instruments for ozone observations, the Scanning Imaging Absorption Spectrometer for Atmospheric Cartography (SCIAMACHY) and the Michelson Interferometer for Passive Atmospheric Sounding (MIPAS), will be launched by the European Space Agency (ESA) on the ENVISAT. The SCIAMACHY is an advanced version of GOME with both nadir and limb capability; it will provide vertical profiles of ozone, nitrogen dioxide, and some other species in the stratosphere and for one or two levels in the troposphere. The MIPAS is a Fourier transform spectrometer for measuring high-resolution gaseous emission spectra at Earth's limb and should have good detection capabilities for ozone (O_3), water (H_2O), HNO_3, nitrous oxide, and CH_4. It was initially developed for the stratosphere but may be able to measure down to 5 km.

CALIBRATION AND VALIDATION

Approaches

Calibration and validation of satellite sensors set the stage on which the integrity of all data subsequently derived will be based. Calibration is the set of prelaunch and on-orbit operations or processes used to determine the relationship between satellite instrument output values and traceable standards. Validation involves evaluating the algorithms required to extract geophysical quantities from calibrated and well-characterized instruments.

As part of an international effort, the Ozone Processing Team (OPT) at the GSFC has committed more than 15 years to refining and validating ozone data from SBUV and TOMS instruments. This effort is being conducted through algorithm improvements and comprehensive studies of pre- and postlaunch calibrations (Hilsenrath et al., 1997). TOMS and SBUV operate on similar principles and employ a common algorithm using the observed Earth's geometrical albedo in the 250 to 380 nm wavelength range. The albedo measurement allows canceling nearly all time-dependent instrument changes common to both the radiance and irradiance measurements. The accuracy of the ozone measurement depends on the accuracy of the prelaunch albedo calibration, the uncertainty in the solar diffuser time-dependent changes, and the application of the algorithm. Other non-canceling parameters, such as instrument linearity, must also be tracked carefully over time.

Consistent prelaunch calibration is essential for traceability when successive instruments of one type are flown. Equally important are accurate absolute calibrations to understand differences among instruments employing different techniques. Calibration involves the use of standards and measurements provided by the National Institute of Standards and Technology (NIST), such as irradiance lamps and integrating sphere radiance targets. Radiance calibrations of the SBUV, TOMS, and GOME instruments have been compared using these standards and thus have common and consistent prelaunch calibrations.

Accurate description of the time-dependent characteristics of an instrument is fundamental to detecting long-term trends. Several methods are now implemented for tracking postlaunch instrument calibration. These methods include on-board calibration systems, algorithmic techniques, and intercomparisons (Hilsenrath et al., 1997). One method of evaluating and calibrating the sensitivity of an aging satellite instrument is to compare observed albedos with those measured from a freshly launched instrument. In accordance with this concept, the U.S. National Plan for Stratospheric Monitoring (NOAA, 1989) called for regular flights of an SBUV instrument (called the SSBUV) on the space shuttle as additional assurance that the drift in the NOAA SBUV/2 instruments would be accurately corrected. SSBUV calibration has been tracked with a precision of 1 percent and has flown eight times from 1989 to 1996.

Pre- and postlaunch calibration is the first step in producing high-quality, long-term environmental data sets from space-based instruments. The second step is data validation. To meet this requirement, a comprehensive program of correlative measurements for in-orbit validation is needed. Ozone column measurements can be validated using ground-based spectrophotometers, and agreement between TOMS and the Dobson network is found to be better than 1 percent (WMO, 1999). Validation of space-based vertical ozone profiles has been done mostly by comparisons with ozonesonde data, but the small number of coincidences is a limiting factor. SAGE-ozonesonde comparisons at 20 to 25 km show no significant difference within an uncertainty range of a few percent but show significant differences above 25 km (potentially from ozonesonde error) and below 20 km (potentially from SAGE error). The SAGE-ozonesonde comparisons further demonstrated no significant instrument drift at 20 to 25 km altitude (less than 0.2 percent yr^{-1}). SAGE II-lidar comparisons for the period 1990 to 1997 and over the altitude range 22.5 to 35 km show an instrument drift of less than 0.5±1 percent yr^{-1} on average.

Challenges

Evidence from past satellite instruments shows that calibration and validation must not be regarded as a one-time effort to be made at the beginning of an instrument flight period but rather an effort that must continue through the lifetime of the instrument. Given the multiplicity of launches involving several countries and agencies, it is important that validation be considered not on a satellite-by-satellite basis but rather as part of an integrated international plan. In this way, the full validation capabilities of all nations' research communities can be applied to the total suite of space-based ozone measurements (Kaye and Miller, 1997).

Validating vertical profiles of ozone is a major challenge, as there are no continuous observational data sets against which to compare the space-based measurements. The SAGE II instrument was initially validated with a dedicated program of ozonesonde measurements (Cunnold et al., 1989; Attmannspacher et al., 1989). Since then, continued validation has had to rely on the regular ozonesonde launch program, which affords few coincidences with the satellite data. Coincidence over a time interval of less than 3 hours appears to be necessary for meaningful comparison (WMO, 1999). Below 20 km altitude, aircraft are a platform of choice for validating space-based observations. The SAGE III Ozone Loss and Validation Experiment (SOLVE), conducted by NASA in the winter 1999-2000 using both the ER-2 and DC-8 aircraft, is the first aircraft mission dedicated to validating space-based ozone data. Additional aircraft missions need to be planned by NASA for validating the sensors of the Chemistry satellite, to be launched in 2002.

EVOLUTION STRATEGY

New Measurement Technologies

Ozone measurements above 25 km altitude are made currently with high reliability by a number of existing spaceborne instruments, including SAGE II, HALOE, SBUV, and MLS. The push for new technology is driven principally by the need to (1) miniaturize and simplify the existing satellite instruments, (2) extend the range of ozone observations to lower altitudes, and (3) measure ozone concurrently with related species to better understand the mechanisms for ozone formation and loss.

The new generation of research instruments, including SAGE III, TES, HIRDLS, and MLS, will provide improved measurement of ozone in the lower stratosphere and the troposphere, along with concurrent measurements of several ancillary species. The TES instrument, designed specifically for tropospheric measurements, has a spectral resolution of 0.025 cm^{-1} in the infrared, corresponding to the width of individual lines in the lower troposphere; this will allow detection of ozone, CO, and water vapor down to the surface. TES has both limb and nadir measurement capabilities to further improve tropospheric detection. The limb observation will provide 2.3 km vertical resolution down to cloud tops, while the nadir observation will provide ~4 km vertical resolution with less likelihood of cloud interference.

Space-based lidars could allow measurement of ozone vertical profiles down to the surface with a vertical resolution of about 2 km. NASA and the Canadian Space Agency (CSA) have jointly proposed the development of a spaceborne lidar system, Ozone Research with Advanced Cooperative Lidar Experiments (ORACLE), to measure ozone profiles in the lower stratosphere and in the troposphere with simultaneous measurements of aerosol and cloud profiles. The Differential Absorption Laser (DIAL) technique will be used to measure ozone profiles and columns in the UV (306 to 318 nm), while direct lidar backscatter returns in the visible or near-infrared will provide the simultaneous aerosol and cloud profile measurements. These techniques have had a long history of application in aircraft studies of tropospheric and stratospheric ozone and aerosols. The transition to space-based applications is made possible by recent advances in tunable solid-state laser technology that have opened the way for compact, efficient, high-power laser systems and advances in composite material and other receiver technologies permitting the development of large-area, lightweight receiver systems.

New Sampling Strategies

Present and planned space-based ozone sensors provide global coverage, but with relatively sparse sampling, i.e., a return time of typically 3 to 7 days on a given tract. This is a significant limitation, considering the large variability of concentrations at extratropical latitudes and in the troposphere. Measurements on a geostationary orbit would provide continuous data over spatial scenes representing one-third of a hemisphere. Nadir observation of O_3, CO, and H_2O with ~2 km vertical resolution down to the surface could be achieved in a geostationary orbit with a Fourier transform spectrometer or a gas-correlation spectrometer.

REFERENCES

Attmannspacher, J. de la Noe, D. De Muer, J. Lenoble, G. Megie, J. Pelon, P. Pruvost, and R. Reiter. 1989. European validation of SAGE II ozone profiles. J. Geophys. Res. 94: 8461-8466.

Cunnold, D.M., W.P. Chu, R.A. Barnes, M.P. McCormick, and R.E. Veiga. 1989. Validation of SAGE II ozone measurements. J. Geophys. Res. 94: 8447-8460.

Hilsenrath, E., P.K. Bhartia, R.P. Cebula, and C.G. Wellemeyer. 1997. Calibration and intercalibration of BUV satellite ozone data. J. Adv. Space Res. 19: 1345-1353.

Integrated Program Office (IPO), National Polar-orbiting Operational Environmental Satellite System (NPOESS). 1996. Integrated Operational Requirements Document (IORD) I. Joint Agency Requirements Group Administrators. 61 pp. + figures.

Intergovernmental Panel on Climate Change (IPCC). 1999. Special Report on Aviation and the Global Atmosphere. Cambridge, U.K.: Cambridge University Press.

Kaye, J.A., and A.J. Miller. 1997. Tropospheric ozone measurements and their use in validation of TOMS and SAGE data products. Earth Observer 9: 31-34.

Logan, J.A. 1994. Trends in the vertical distribution of ozone: An analysis of ozonesonde data. J. Geophys. Res. 99: 25553-25585.

Logan, J.A. 1999. An analysis of ozonesonde data for the lower stratosphere: Recommendations for testing models. J. Geophys. Res. 104: 16151-16170.

Logan, J.A., I.A. Megretskaia, A.J. Miller, G.C. Tiao, D. Choi, L. Zhang, L. Bishop, R. Stolarski, G.J. Labow, S.M. Hollandsworth, G.E. Bodeker, H. Claude, D. DeMuer, J.B. Kerr, D.W. Tarasick, S.J. Oltmans, B. Johnson, F. Schmidlin, J. Staehelin, P. Viatte, and O. Uchino. 1999. Trends in the vertical distribution of ozone: A comparison of two analyses of ozonesonde data. J. Geophys. Res. 104: 26373-26399.

National Oceanic and Atmospheric Administration (NOAA). 1989. National Plan for Stratospheric Monitoring and Early Detection of Change, 1988-1997. FCM-P17-1989. U.S. Department of Commerce, July.

Weatherhead, E.C., G.C. Reinsel, G.C. Tiao, X.L. Meng, D.S. Choi, W.K. Cheang, T. Keller, J. DeLuisi, D.J. Wuebbles, J.B. Kerr, A.J. Miller, S.J. Oltmans, and J.E. Frederick. 1998. Factors affecting the detection of trends: Statistical considerations and applications to environmental data. J. Geophys. Res. 103: 17149-17161.

World Meteorological Organization (WMO). 1999. Scientific Assessment of Ozone Depletion: 1998. Geneva: WMO.

9

Earth Radiation Budget

INTRODUCTION

The Earth radiation budget (ERB) is a combination of the broadband fluxes of solar radiation reflected by Earth and the fluxes of longwave radiation absorbed and emitted by Earth and its atmosphere. The net radiation N at the top of the atmosphere (TOA) may be defined in terms of three quantities (Hartmann, 1994):

$$N = S(1 - \alpha) - I$$

where S is the solar insolation, α is the planetary albedo, and I is the outgoing longwave radiation emitted by Earth to space. The goal of ERB measurement programs is to provide observations of the space and time distributions of both α and I and usually S as well (Wielicki et al., 1995).

The radiation budget is a critical component of the energy budget of Earth's climate system and is thus fundamental to the study of climate. The variation of the radiation budget with latitude is the principal force driving the transport of heat from low to high latitudes via the circulations of the atmosphere and oceans. Changes in the radiation budget induced by increasing concentrations of greenhouse gases and aerosols and changes in the solar insolation at the top of the atmosphere define the radiative forcings of climate. Modulations of the radiation budget associated with changing surface and atmospheric conditions, including clouds, give rise to significant climate feedbacks that are considered to be one of the most uncertain aspects in our understanding of climate and climatic change (IPCC, 1995).

ERB measurements seek to contribute to two key scientific goals: (1) the determination of how long- and shortwave fluxes are distributed in space and how they vary in time and (2) the development of a quantitative understanding of the links between the radiation budget and the properties of the atmosphere and surface that define that budget.

The committee's assessment of the current ERB observation systems and National Polar-orbiting Operational Environmental Satellite System (NPOESS) plans for future long-term measurements is presented in Box 9.1.

Box 9.1
Findings

Present satellite Earth radiation budget (ERB) measurement programs have provided valuable observations that have advanced our understanding of the two science issues described in the introduction to this chapter, and further advances in this understanding are expected in the Earth Observing System (EOS) and post-EOS eras. However, understanding of the radiation budget of the planet is still largely confined to top-of-the-atmosphere (TOA) fluxes; we have not made significant progress in achieving the science goals stated above. Although TOA information is important for a number of reasons, it is unable to give direct insight into processes that influence the radiation budget of the atmosphere and surface.

The Clouds and the Earth's Radiant Energy System (CERES) developed for EOS (Wielicki et al., 1996) can meet the TOA requirements defined in the NOAA IORD-1 climate requirements document (IPO NPOESS, 1996). Surface flux requirements, however, cannot be met with these measurements alone and require significant ancillary information. The requirements for some of this additional information (such as cloud base) cannot be met entirely with planned National Polar-orbiting Operational Environmental Satellite System (NPOESS) observations, although the issue of how much cloud base information is contained in the NPOESS observing system is a topic of ongoing research.

Future ERB instruments should not be simply a copy of CERES but should have enhanced capabilities that CERES does not provide. The next advance in ERB measurements must come in the direction of providing a better and more direct way of determining the radiation budget at the surface, within the atmosphere, and at the top of the atmosphere. The challenge is to determine the most appropriate enhancement to achieve this goal (one example is an enhanced spectral flux measurement capability).

There is an ongoing debate on the climatic value of continuous ERB measurements. Closure on this debate is needed, and the extent to which the data expected from EOS will advance progress toward the stated goals of an ERB measurement program will have to be assessed. Therefore current planning activities should not preclude the incorporation of CERES-like or ideally enhanced ERB observations as part of the NPOESS climate-observing strategy.

This should include the NPOESS Preparatory Project Pathfinder mission being planned to bridge the gap between the end of the first EOS missions in 2006 and the start of NPOESS in 2009.

The next steps in ERB observations might include:

1. *A calibrated spectral imager to fly alongside a CERES-like instrument.* Imager data are required for scene identification and to define the appropriate angular model used to retrieve flux information. These data must be calibrated and contain more capable cloud channels beyond those channels used for ocean color. Current AVHRR-like channels are not sufficient for this purpose, but the channels of MODIS and those planned for VIIRS would provide an adequate means for scene identification.

2. *Sampling requires measurements from multiple satellites.* A minimum of two orbits is required to sample the diurnal cycle, but three are preferable.

3. *A measurement strategy that provides a more direct way of determining the radiation budget at the surface and within the atmosphere.* The challenge is to determine the most appropriate observing strategy to achieve this goal (such as more spectral flux measurement capability).

4. *Correlative measurements of the principal atmospheric constituents governing the distribution and variability of the fluxes measured at the TOA.* This ideally should include information on cloud water and ice path as well as aerosol optical depth.

RADIATION BUDGET IN THE SATELLITE ERA

Two ERB activities have contributed significantly to observations of the radiation budget: the Earth Radiation Budget Experiment (ERBE) (Barkstrom et al., 1989) and the Clouds and the Earth's Radiant Energy System (CERES) (Wielicki et al., 1996). Progress in measurements of the ERB from space can be summarized in terms of three observing eras: before 1984, 1984 to 1997, and after 1997.

The Pre-ERBE Era (Before 1984)

Before ERBE, the radiation budget was determined using a variety of data sets collected from different experimental satellite programs, notably from the Nimbus series of experimental satellites (House, 1985). These early contributions significantly improved our understanding of ERB. For example, the earliest satellite observations led to a downward revision of Earth's albedo, from the pre-satellite-era value of 50 percent to near the current value of 30 to 31 percent (Vonder Haar and Suomi, 1971). Most of the observations before 1984 came from flat-plate sensors that collected radiation over a broad region of Earth (at an approximate resolution of 10 degrees latitude/longitude). Estimates of the effects of clouds on the radiation budget were attempted using these data, but the results were hampered by the coarse resolution of the measurements. Errors attached to these early flux data are difficult to quantify, and the best estimates are approximately 10 Wm^{-2} (Ellis et al., 1978).

The ERBE Era (1984 to 1997)

ERBE flew both wide-field-of-view, flat-plate radiometers and the narrow-field-of-view scanning radiometer (Barkstrom et al., 1989). Attention has focused on analyses of measurements from the scanner, which provided data for approximately 6 years before failing in 1990. The improved spatial resolution of flux data achieved with this scanner is perhaps the most important advance offered by ERBE, because it led to better estimates of the difference between cloudy and clear-sky fluxes and thus a better estimate of the effect of clouds on the ERB (Harrison et al., 1990). The usefulness of this information for testing global climate models is now well documented. Monthly mean ERBE flux errors are estimated at 5 to 10 Wm^{-2}.

The CERES Era (post-1997)

The first of the CERES experiments was launched in 1997 on the Tropical Rainfall Measuring Mission (TRMM) satellite (Wielicki et al., 1996). The measurement approach of CERES is ostensibly the same as that of ERBE, with advances in the improved angular sampling and cloud classification that reduce the monthly mean flux error to 1 or 2 Wm^{-2}. The TRMM CERES instrument has not been operational since August 1998, owing to a technical problem that should not affect the performance of the CERES instruments on Earth Observing System (EOS) AM and EOS PM.

OBSERVING STRATEGY

There are two approaches to the satellite measurement of radiative fluxes at the TOA. One approach uses uncalibrated operational satellite data; the second approach, which is more direct, uses calibrated, spectrally integrated data such as that provided by ERBE and CERES to determine these fluxes. This more direct approach relies on measurements of broadband (i.e., spectrally integrated) radiances to obtain fluxes and requires angular correction factors (described below). The less direct approach was developed largely out of necessity and was designed to provide information when none was available from other sources (Minnis et al., 1991). The data from the less direct approach are used to fill gaps in time series of radiation budget data and to improve on the less-than-ideal sampling characteristics of existing data.

The indirect approach employs the narrow (spectral) band radiance measurements available from scanning operational sensors to determine fluxes. These data are usually uncalibrated and the method of conversion requires two corrections: one angular correction to convert radiances to fluxes and a second broadband correction to convert spectral information to broadband information. Estimates of solar fluxes using this method are highly uncertain; their value is dubious and their quality debatable. By contrast, the estimation of longwave fluxes from spectral radiance data from the High-resolution Infrared Sounder (HIRS) has been shown to be more reliable (Ellingson et al., 1994).

It should be noted that the instrument capabilities and measurement strategies differ substantially between shortwave and longwave ERB observations. For example,

- Highly calibrated spectra measurements are much easier to obtain for longwave radiation. So far, only the broadband shortwave ERB measurements have demonstrated sufficient calibration accuracy and stability.
- Averages of TOA fluxes and surface fluxes, at climate time and space scales, are expected to be closely related for shortwave fluxes (approximately linear) but only weakly related for longwave fluxes.
- Time sampling and angular sampling are more difficult for shortwave than for longwave ERB, as is clear from the error analysis in Table 9.1.

These differences suggest that different strategies may be needed for shortwave and longwave flux measurements to improve on the CERES capability.

TABLE 9.1 Error Analysis of ERB Measurements of Flux, Shortwave vs. Longwave

	Monthly Average Regional 5-yr Trend $S_0 = 348$ Wm^{-2}		Monthly Zonal Average Equator-Pole Radiation Difference (Wm^{-2})		Monthly Average Regional, 1 Standard Deviation $S_0 = 348$ Wm^{-2}		Instantaneous Pixel, 1 Standard Deviation $S_0 = 1,000$ Wm^{-2}	
	ERBE[a]	CERES[b]	ERBE	CERES	ERBE	CERES	ERBE	CERES
Shortwave radiation								
Calibration	2.0	1.0	0.2	0.1	2.1	1.0	6.0	3.0
Angle sampling	0.0	0.0	12.0	4.0	3.3	1.1	37.5	12.5
Time sampling	0.0	0.0	2.9	1.4	3.9	1.9	0.0	0.0
Space sampling	0.3	0.3	0.0	0.0	0.3	0.3	0.0	0.0
Total error	2.0	1.1	12.3	4.3	5.5	2.5	38.0	12.9
Longwave radiation								
Calibration	2.4	1.2	2.6	1.3	2.4	1.2	2.4	1.2
Angle sampling	0.0	0.0	2.0	0.7	1.6	0.5	12.5	4.2
Time sampling	0.0	0.0	0.6	0.6	1.3	1.3	0.0	0.0
Space sampling	0.2	0.2	0.0	0.0	0.2	0.2	0.0	0.0
Total error	2.4	1.2	3.3	1.6	3.2	1.9	12.7	4.3
Net radiation								
Calibration	3.1	1.6	2.6	1.3	3.2	1.6	6.5	3.2
Angle sampling	0.0	0.0	12.2	4.1	3.7	1.2	39.5	13.2
Time sampling	0.0	0.0	2.9	1.6	4.1	2.3	0.0	0.0
Space sampling	0.4	0.4	0.0	0.0	0.4	0.4	0.0	0.0
Total	3.1	1.6	12.8	4.5	6.4	3.1	40.1	13.6
Science requirement[c]	2 to 5	<1	10	1 to 3	10	2 to 5	None	10

NOTE: S_0 is the global, annual mean solar insolation at the top of the atmosphere.
[a]ERBE: Crosstrack scanner only, two satellites, 2.5° latitude/longitude regions.
[b]CERES: Crosstrack and rotating azimuth scanners, MODIS, three satellites, 1.25° regions.
[c]As specified by CERES.

Flux Retrievals

TOA fluxes are retrieved from radiance data using precomputed angular models that convert the measured radiances L to hemispheric fluxes F according to

$$F = \pi L R^{-1}$$

where R is the Angular Directional Model (ADM), which accounts for the anisotropy of the radiation field. The best way of obtaining broadband fluxes F is to provide calibrated broadband measurements of L. While it is important to maintain a high level of accuracy in the calibration of L, the largest sources of error in retrieved fluxes relate to the uncertainty in the angular models (see Table 9.1). This uncertainty arises partly from the lack of full angular sampling and partly from the complexity of the dependence of R on the properties of the scene, in particular the dependence on cloud properties (such as optical properties and cloud volume). ERBE uses 12 observationally derived ADM types based loosely on cloud volume (Wielicki and Green, 1989).

One of the reasons CERES is able to reduce the error associated with the retrieval of flux is the better resolution of the ADMs obtained from a broader angular sample and the better cloud property information from related measurements. The additional information about the scene is derived using data from another sensor, such as the Visible and Infrared Imaging System (VIIRS) on TRMM or the Moderate-resolution Imaging Spectroradiometer (MODIS) on AM and PM. In principle, classifying the ADMs in terms of a larger class of properties provides a way of defining the appropriate ADM more accurately. The number of ADM categories proposed by CERES is more than an order of magnitude larger than the 12 categories used for ERBE. The CERES team estimates that the better-resolved ADMs, derived from the better sampling and classification of CERES, will reduce the ADMs' contribution to error by a factor of approximately 3 (see Table 9.1).

Sampling Characteristics

The second important component of the TOA flux observing system, and one common to many climate measurement problems, is the nature of the sampling of the measurements. Undersampling of fluxes in space and time is responsible for much of the error budget of fluxes (see Table 9.1). These errors arise from the limitations associated with the asynoptic nature of the sampling of polar-orbiting satellites, which observe different locales at different instants in time. This is a problem for any observing system on polar-orbiting satellites, because space and time behaviors are mixed. Asynoptic sampling biases on time mean properties are a particular problem for those properties that undergo marked diurnal variation, typical of many climate processes, particularly those related to hydrological processes (clouds and precipitation and the resulting radiative fluxes).

Although the systematic error associated with undersampling the diurnal cycle is most serious for polar-orbiting measurements, in which diurnal variability is indistinguishable from the time-mean, it surfaces even in processing measurements. It occurs because the diurnal cycle is not perfectly repeatable, it is spatially coherent and therefore difficult to remove, and it is sampled too slowly to be truly resolved in such observations. As a result, time-mean behavior is analyzed by undersampled diurnal variability (as is low-frequency behavior in general), making errors in the two closely related (Salby and Callaghan, 1996).

Three distinct approaches have been proposed to address this problem:

1. The most obvious is to use a combination of observations from multiple satellites that provide adequate sampling of the diurnal cycle. In the CERES era, this will include a combination of measurements from the AM and PM satellites as well as from the precessing TRMM satellite. Simulations of the sampling errors in TOA fluxes that are due to undersampling by a single satellite and by combinations of all three satellites demonstrate how these errors can be systematically reduced by adding temporal sampling provided by measurements from two and three satellites.

2. The second approach is to use geostationary data to provide the characteristic diurnal variability of spectral radiances (obtained from geostationary imagers) and apply this variability to a model of the diurnal cycle to correct

sampling biases in the broadband fluxes (Brooks and Minnis, 1984). A variation of this strategy has been adopted in analyzing CERES data collected under the TRMM. In this case, narrowband, 3-hourly geostationary data are used to provide the magnitude of the diurnal cycle, while a single broadband sensor on a low Earth orbit provides the broadband reference field. The diurnal correction is treated as a second-order correction, and so it is less critically dependent on the calibration of the geostationary data or the broadband/narrowband relationship. The accuracy of this approach will be demonstrated by comparing all three CERES measurements (TRMM, EOS-AM, EOS-PM) to results for any one satellite plus geostationary data.

3. The third approach is to fly a radiation budget instrument on the geostationary satellite, thereby providing proper sampling of the space-time variations of the radiation budget and eliminating the diurnal sample error. The first Geostationary Earth Radiation Budget (GERB) instrument is to be launched as part of the payload of the MeteoSat Second Generation (MSG) in 2001. It will provide an opportunity to assess the nature of sampling errors, at least over the region observed by the MSG 1 field of view, and to test the sampling error corrections developed for the other data sets. There are a number of complications associated with geostationary observations of the ERB. First, the observations are not global. Second, unlike low-Earth-orbiting satellites, the geostationary satellite view limits the climate observations for any given region of Earth to a single viewing zenith angle. This limitation causes an even larger dependence on angular model corrections and will increase the angular sampling errors shown in Table 9.1. As a result, angular sampling errors will be aliased into spatial patterns of the radiation field. The magnitude of these errors has yet to be quantified, but it can be determined once sufficient overlapping CERES and GERB data have been compared.

Key Elements of the ERB Observing System

The key elements of an ERB observing system are the following:

- Broadband or narrowband radiometers (preferably scanning) providing calibrated radiance data;
- A capability for coincident scene identification, including an ability to determine cloud properties, which requires a calibrated imaging radiometer;
- A data reprocessing strategy applying ADMs determined from accumulated observations. For instance, CERES flux estimates will come from reprocessing data after the CERES ADM statistics are categorized and accumulated. It is not known whether these ADMs constructed from CERES are sufficient for application to other (future) data. Preliminary CERES data sets using ERBE ADMs will be provided for initial releases of data.

The factors that contribute to the error budget of the ERB observing system, as suggested in Table 9.1, vary according to the space-time averaging adopted in the data. For example, at instantaneous time scales, angular sampling errors dominate the radiation budget measurements. For daily regional average estimates, time sampling dominates. For monthly regional means, there is a rough balance between time sampling, angle sampling, and calibration uncertainty. For interannual zonal or regional means, calibration uncertainty dominates. The ability to monitor regional and zonal climate change therefore depends most critically on calibration stability for overlapping measurements and on absolute accuracy for nonoverlapping measurements.

CALIBRATION AND VALIDATION STRATEGIES

Maintenance of a precise and accurate calibration is an important part of the ERB measurement strategy. The goal of CERES is 0.5 percent for longwave fluxes and 1 percent for shortwave fluxes. A key to achieving these calibration goals is the combination of prelaunch calibration and in-orbit calibrations performed approximately every 2 weeks. The prelaunch calibration is performed at a special calibration facility that provides a fundamental traceability of the data to reference standards. In-orbit calibration uses a combination of blackbody sources, deep-space views, and views of the Sun through a diffusing mirror, although there are significant errors associated with the last.

The current CERES absolute accuracies stated above translate to an absolute accuracy of between 1 and 3 Wm^{-2} in net flux, similar to the magnitude of greenhouse-gas-related climate forcings. The calibration stability of the CERES instrument on TRMM, however, is better than 0.25 Wm^{-2} for shortwave and longwave. This stability offers the first demonstrated ability to achieve direct measurements of climate change anomalies with a precision significantly higher than that projected for currently known climate forcings.

OPPORTUNITIES

The GERB sensor on MSG 1 is expected to be launched in 2001 and continued on both MSG 2 and MSG 3. Unfortunately, current U.S. satellite plans do not include TOA flux measurements beyond EOS. It would be possible, however, to add a radiation budget instrument to the NPOESS Preparatory Project (NPP) Pathfinder mission being planned to bridge the gap between the end of the first EOS missions in 2006 and the start of NPOESS in 2009. Ideally for ERB climate monitoring, the NPP mission would fly with a broadband radiation measurement and calibrated cloud imager, in a 1:30 p.m. Sun-synchronous orbit to overlap EOS PM and the NPOESS 1:30 p.m. orbit. In this case, all three missions (EOS PM, NPP, and NPOESS) would provide the necessary broadband radiation and cloud imager data to continue the climate time series.

There is a concern that radiation budget measurements on NPOESS may have a lower priority when it comes to funding than weather measurements. In this case, the ERBE experience might be repeated: funding cuts or cost overruns could eliminate the radiation budget measurement from the NPOESS platform. In the committee's view, NOAA and NASA should develop a strategy to avoid this: NASA flies the prototypes, but NOAA cannot afford to continue the climate time series. From NOAA's perspective, this is a matter of priorities: weather is a higher priority than climate. A solution to this might be for NASA to fund the climate-oriented broadband radiometer, while NOAA provides the spacecraft, calibrated imager, and launch that are required to support the higher-priority weather mission.

LIMITATIONS AND THE EVOLUTION STRATEGY

The emphasis of existing radiation budget measurements has largely shifted toward reducing the error characteristics of the measurements. As noted above, this involves three phases of activity:

1. Maintaining high calibration accuracy,
2. Improving the accuracy of the ADMs by improving the angular sampling and scene classification, and
3. Improving the time and space sampling.

Current ERB satellite measuring programs provide useful information about the TOA fluxes, but this information has its limitations, two of which follow:

1. Progress toward confirming our estimates of radiative forcing of climate requires careful monitoring of the radiation budget. These forcings are small, and except for episodic events like volcanic eruptions (Minnis et al., 1993), the radiative forcings associated with these relatively short periods of observation have been too small to be detectable by present TOA broadband-flux observing systems. Current CERES has a precision that can detect radiative forcing. It is perhaps unrealistic to expect significant improvements in the accuracy of ERB measurements beyond those claimed by CERES. Climate responses such as temperature and humidity profiles, however, have a more discernible and easily detectable spectral signature in the thermal infrared (Goody et al., 1995). Achieving highly calibrated, high-spectral-resolution data, however, may require reduced spatial sampling, such as a nadir-only view and a field-of-view size of roughly 100 km. For use in climate monitoring, further quantification is needed to study the trade-off between the increased signal gained with spectral resolution and the increased noise caused by limited spatial sampling. It may be necessary to combine highly calibrated instruments.

2. Developing a quantitative understanding of the links between the radiation budget and the properties of the atmosphere that define these fluxes continues to be an important motivation for research on and measurement of the radiation budget (Wielicki et al., 1995). Progress on this topic continues to be elusive, however, owing in part to the limitations of existing TOA information. The nature of these limitations is exemplified in studies of the effects of clouds on TOA fluxes. It is well documented that the net effect of clouds on TOA fluxes is a balance between changes in solar fluxes associated with the larger albedo of clouds and changes in longwave emission to space associated with changes in cloud top altitude. These changes produce reciprocal effects that approximately balance at the top of the atmosphere but not within the atmosphere or at the surface. With NASA's recent selection of the CloudSat and PICASSO-CENA ESSP missions for launch, key measurements will begin to address this major limitation.

The next step in ERB measurements requires a more capable radiation budget observing system than currently exists. This new system should provide observations that will (1) establish a more direct link between the observed fluxes and the parameters that affect these fluxes and (2) provide an improved ability to measure the flux distributions within the atmosphere and at the surface. This will only be partially met by the coincident measurements of relevant properties such as cloud information and ERB fluxes expected in the EOS era.

REFERENCES

Barkstrom, B., E.F. Harrison, G.L. Smith, R.N. Green, J. Kibler, R. Cess, and the ERBE Science Team. 1989. Earth Radiation Budget Experiment (ERBE) archival and April 1985 results. Bull. Am. Meteorol. Soc. 70: 1254-1262.

Brooks, David R., and Patrick Minnis. 1984. Simulation of the Earth's monthly average regional radiation balance derived from satellite measurements. Journal of Climate and Applied Meteorology 23(3): 392-403.

Ellingson, R., H-T. Lee, D. Yanuk, and A. Gruber. 1994. Validation of a technique for estimating outgoing long-wave radiation from HIRS radiance observations. J. Atmos. Ocean. Technol. 11: 357-365.

Ellis, J.S., T.H. Vonder Haar, S. Levitusand, and A.H. Oort. 1978. The annual variation of the global heat balance of Earth. J. Geophys. Res. 84: 1958-1962.

Goody, R.M., R. Haskins, W. Abdou, and L. Chen. 1995. Detection of climate forcing using emission spectra. Issledovaniye Zemli is Kosmosa (Earth Research from Space) 5: 22-23.

Harrison, E.F., P. Minnis, B. Barkstrom, V. Ramanathan, R.E.D. Cess, and G.G. Gibsoin. 1990. Seasonal variation of cloud radiative forcing derived from the ERBE. J. Geophys. Res. 95: 18667-18703.

Hartmann, D. 1994. Global Physical Climatology. San Diego: Academic Press, 408 pp.

House, F.B. 1985. Observing the earth radiation budget from satellites: past, present and a look to the future. Adv. Space Res. 5: 89-98.

Integrated Program Office (IPO), National Polar-orbiting Operational Environmental Satellite System (NPOESS). 1996. Integrated Operational Requirements Document (IORD) I. Joint Agency Requirements Group Administrators. 61 pp. + figures.

International Panel on Climate Change (IPCC). 1995. The Science of Climate Change, Houghton et al. (eds.), Cambridge, U.K.: Cambridge University Press, 572 pp.

Minnis, P., D.F. Young, and E.F. Harrison. 1991. Examination of the relationship between outgoing infrared window and total longwave fluxes using satellite data. J. Climate 4: 1114-1133.

Minnis, P., E.F. Harrison, L.L. Stowe, G.G. Gibson, F.M. Denn, D.R. Doelling, and W.L. Smith, Jr. 1993. Radiative climate forcing by Mt. Pinatubo eruption. Science 259: 1411-1415.

Salby, M.L., and P. Callaghan. 1996. Sampling error in climate properties derived from satellite measurements: relationship to diurnal variability. J. Climate 10: 18-36.

Vonder Haar, T.H., and V. Suomi. 1971. Measurements of the earth's radiation budget from satellites during a 5-year period: I: Extended time and space means. J. Atmos. Sci. 28: 305-314.

Wielicki, B., and R.N. Green. 1989. Cloud identification for ERBE radiative flux retrieval. J. Appl. Meteorol. 28: 1133-1146.

Wielicki, B., B. Barkstrom, E.F. Harrison, R.B. Lee, G.L. Smith, and J. Cooper. 1996. Clouds and the earth's radiant energy system (CERES): an earth observing system experiment. Bull. Am. Meteorol. Soc. 77: 853-867.

Wielicki, B., R.D. Cess, M.D. King, D.A. Randall, and E.F. Harrison. 1995. Mission to planet earth: role of clouds and radiation in climate. Bull. Am. Meteorol. Soc. 76: 2125-2153.

10

Issues, Challenges, and Recommendations

COMMON ISSUES

The committee's review of the eight selected measurement sets in terms of their value for climate research illuminated critical issues to be addressed by an integrated satellite observing system for climate research and monitoring. Specific findings are summarized in Chapters 2 through 9. Here, the committee highlights the technical and programmatic issues that are common across all measurements. These include both technical issues related to the need to ensure data continuity and interoperability as well as programmatic issues related to the integration of research with operational missions.

Monitoring Trends and Understanding Processes

• **Need for a comprehensive long-term strategy.** Systems for observing climate-related processes must be part of a comprehensive, wide-ranging, long-term strategy. Monitoring over long time periods is essential to detecting trends, as discussed in Chapter 2 with respect to the importance of trends in radiance data records, in Chapter 4 dealing with trends in land-cover change, and also in Chapter 8 concerning trends in ozone destruction. Long-term monitoring is also necessary to understand critical *processes* that are characterized by low-frequency variability. Process studies focusing on specific regions or issues of scientific interest are another critical element. Because changes on a wide range of time and space scales affect Earth, it is not possible to determine a priori and with certainty the types of observations that should be made and the appropriate sampling strategy. An observing system may very well reveal unexpected phenomena such as the large-scale, low-frequency El Niño/Southern Oscillation, as discussed in Chapter 3 on sea surface temperature, and scientific opportunities are lost if the observing strategy cannot adapt accordingly.

• **Desirability of multiple measurements of the same variable using different techniques.** Many of the processes of interest for climate research and monitoring can be observed using multiple techniques. If such measurements corroborate results, revealing similar patterns and trends, then confidence in the overall quality of the data increases. Similarly, if the measurements are in conflict, then this information may suggest problems in data quality or newly emerging science questions that need resolution. The advantages of multiple measurements obtained with a variety of techniques are discussed for the study of atmospheric temperature and moisture soundings (Chapter 2), sea surface temperature (Chapter 3), land cover (Chapter 4), aerosols (Chapter 7), and Earth's radiation budget (Chapter 9).

- **Diversity of satellite observations and sampling strategies and support for ground-based networks.** An effective climate observing system depends on diverse satellite observations and sampling strategies, including measurements made by ground-based networks. The measurement tools must be responsive to the requirements imposed by the variables to be observed. While NPOESS and EOS have focused primarily on polar-orbiting satellites, satellite observations from other orbits (low inclination, geostationary) are also necessary (see Chapters 2, 3, 8, and 9). Some critical sensors will be flown by international partners, and so it will be important to integrate scientific findings on a more global basis (as discussed in Chapters 1, 3, 4, 5, 7 and 8).

Surface- or ground-based networks to support process studies are also required, as discussed in Chapter 4, 5, and 7; concern for the neglect of these is expressed in Chapter 4. High-resolution measurements in both time and space (as discussed in Chapters 4, 5, 6, and 7) are a critical element of many process studies. As the observing systems improve, it is likely that undersampling (in time and space) will come to dominate the errors in the system (see Chapters 1, 2, 7, and 8). Innovative strategies, such as the use of satellite constellations, may be important for some applications, as expressed in Chapter 3.

- **Preserving the quality of data acquired in a series of measurements.** It takes a special effort to preserve the quality of data acquired with different satellite systems and sensors, so that valid comparisons can be made over an entire set of observations. There are few examples of continuous data records based on satellite measurements where data quality is consistent across changes in sensors, even when copies of the sensor design are used. Sensor characterization and an effective, ongoing program of sensor calibration and validation are essential in order to separate the effects of changes in the Earth system from effects owing to changes in the observing system. Providing for overlap across successive sensors is critical, especially given the regular insertion of new technology driven by the need to reduce costs and/or improve performance. Concern about preventing data gaps, ensuring data continuity, and developing overlap strategies is expressed in Chapters 3, 4, 5, 7, 8, and 9. Data systems should be designed to meet the needs for periodic reprocessing of the entire data set. An aggressive, science-driven program to ensure long-term data quality and continuity is very important.

- **The role of data analysis and reprocessing.** Improvements in understanding will come from continued, thorough analysis of new and ongoing observations. An active program of data analysis and reprocessing will add value to existing data sets by expanding the purposes for which they can be used and will enable development of new algorithms and new data products. Such analysis will also provide the basis for investment in new technologies for needed improvements such as innovative sensors.

- **Technology development and improved measurement capabilities.** New sensors are needed to reduce costs and to improve measurement capabilities. For example, higher-resolution sensors could help resolve some of the still open questions discussed in Chapters 4, 5, and 7. Moreover, because all critical climate-related variables may not yet have been identified, and some, such as soil moisture (Chapter 6), cannot yet be measured effectively from space, continued technology and science investment is required. By coordinating its technology efforts with those of IPO/NPOESS, NASA could address these issues and help provide increased capabilities for the operational meteorological system.

Status of NASA and IPO/NPOESS Integration

- **Division of responsibility in the integration of research and operational missions.** Climate research and monitoring raise issues that transcend the capabilities of any single federal agency. There is currently no effective structure in place in the federal government that can address such multiagency issues as the balance between satellite and ground observations, long-term and exploratory missions, and research and operational needs. The committee concurs with the several recent NRC reports that have expressed concern over the lack of overall authority and accountability, the division of responsibility, and the lack of progress in achieving a long-term climate observing system (see, for example, NRC 1998a,b and 1999a,b; see also footnote 3 in Chapter 1). The challenges in integration of NASA/ESE research satellite missions and IPO/NPOESS operational satellite missions highlight the critical issues.

- **Adequacy of NPOESS environmental data records for climate research.** The current emphasis by IPO/NPOESS on environmental data records does not ensure development of a system that will meet the requirements for climate research. The EDR process established by IPO/NPOESS supports the primary operational needs of DOD and NOAA but was not intended to yield instrument specifications that meet climate research requirements (see Chapters 1, 4, 5, 7, and 8. For example, many climate research studies require access to unprocessed sensor-level data, whereas the EDR approach focuses on the final data products. In many cases, the current EDRs are not completely specified, and in some, they are not adequate for climate research because they do not, for example, specify measurement stability and longevity (see Chapters 4, 5, 7, and 8).
- **The NASA/ESE approach to long-term measurements.** The current NASA/ESE approach relies on a strategy of systematic measurements that may be integrated into NPOESS with no assurances that this integration will be successful. Long-term observations are essential for climate studies, yet NASA's new EOS plan focuses on short-term (3- to 5-year) missions. For NASA to be able to pursue a science-based strategy that leverages NPOESS capabilities where possible, the agency will probably also have to fly complementary missions and to collect some specialized data sets (see Chapter 1 for a discussion of NASA's and NOAA's approaches to long-term measurements).
- **A systemwide analysis of research and operational requirements is not currently in place.** Such an analysis is necessary to ensure that scientific and programmatic requirements are met on a continuing basis. The NPOESS Preparatory Project is an encouraging step in addressing the need to maintain continuity of critical data sets between the end of the EOS platforms and the launch of the first NPOESS platforms. However, there is no assurance that the NPP experience will be repeated for other important data sets. For example, the NPOESS IPO is planning to derive vector winds from passive microwave data; however, launch of the planned Navy Windsat satellite may be necessary to prove the technique. Moreover, there is no assurance that the Navy will continue the Windsat program, and NASA support for follow-ons to its scatterometers is also not assured.
- **Development of sustainable instrumentation.** There is no indication, at present, that as measurement capabilities are integrated, the development of sustainable instrumentation will receive the necessary attention. Research instruments often are complex and are intended to make ambitious state-of-the-art measurements. Operational sensors generally should be less expensive to build and operate and above all else must be reliable, requirements that should be balanced with those for climate research, coupled with appropriate stability and longevity requirements.
- **Prioritizing and establishing an observing strategy.** The climate research community has not yet prioritized critical data sets or developed an overall national observing strategy, including algorithm development, sensor calibration and validation, ground observations, and new technology. Priorities should reflect scientific need, while recognizing technological, fiscal, and programmatic constraints. Other important aspects of such a strategy will be periodic evaluation and readjustment of specific mechanisms for transferring data sets from research to operations. Articulation of a long-term context, spanning as much as a century or more, will be paramount. Without such a commitment, a climate observing policy will be meaningless.

THE CHALLENGES OF SPACE-BASED CLIMATE RESEARCH

There are many challenges associated with the development of high-quality, long-term, satellite-based time series suitable for detection of climate change as well as for characterization of climate-related processes. In addition to addressing technical issues, satellite-based climate observing strategies must take into account conflicting requirements and differing agency cultures. These well-known structural problems and issues are not unique to satellite missions; similar problems confront ground-based climate observing strategies (NRC, 1999a). Nevertheless, despite such challenges, there is now an opportunity to make progress toward developing satellite-based observing systems that, while not optimum, will support climate research.

- Climate research and monitoring require a blend of short-term, focused measurements as well as systematic, long-term measurements (see Chapter 1).
- Various problems have precipitated an impasse that inhibits researchers' ability to study and monitor

climate change from either satellites or in situ systems. Measurements often are not made with sufficient accuracy or precision (see Chapters 2, 3, and 8), changes in instrumentation lead to inconsistent data records, and the records are sporadic. Data archives are not designed to facilitate large-scale reprocessing of data, and pricing policies discourage large-volume data extraction. The political and programmatic pressures for short-term returns (both in terms of science and protection of life and property) lead to short changing of the chronic problems of climate change and climate variability in favor of the acute problems of storms, earthquakes, and other severe events.[1]

- Current federal divisions of responsibility constitute a structural impediment to the integration of operational and research components for Earth observation. Overcoming these structural obstacles will require the initiative of the Executive Branch of the federal government to build a broad consensus and then assign clear responsibilities and authority. The outcome of this effort would be a strategy that balances the interests and needs of various constituencies. Key stakeholders include not only the climate research community and the operational agencies, but also the broader Earth science community and other public and private entities that stand to benefit from a strong national program of Earth observations and research. These diverse stakeholders need to be brought together to achieve goals that transcend those of the individual entities. Reaching agreement on how to resolve issues will require compromise and collaboration, rather than mandates.

- There is, at present, no effective forum within the government for making trade-offs and weighing options with regard to short-term forecasting systems for operational needs, long-term, systematic observing systems, and process studies needed for climate research. Research agencies such as NASA are wary of any commitment to maintaining long-term operational satellite monitoring systems that may prevent them from pursuing new technologies and new research directions. Operational agencies such as NOAA and DOD are wary of committing to long-term observation requirements that may impose significant new costs on an already-cost-constrained system. The Earth science community, while recognizing the need for long-term observations for some climate processes, is wary of relinquishing responsibility for and oversight of instrumentation to an operational agency that may find an "instrument facility" approach necessary to manage multiple demands and the long-term stability and performance of the sensor system. Achieving progress will require the initiative of the federal government, with active planning and continuing involvement by the climate research community.

- Various agency configurations have been proposed as solutions to this dilemma. For example, the climate-observing mandate might be given to an operational agency such as NOAA. At the very least, NRC panels have suggested that the Executive Branch establish an office to manage a climate observing system (NRC, 1998a, 1999b). However, climate research and monitoring require a culture and a program structure far different from those currently in place in NOAA. Moreover, there is still a strong research component that must be maintained; the collection of many climate measurements is simply not ready to be placed under operational control, and new measurements and technologies aimed at augmenting or improving the operational measurement suite ought to continue to be developed.

An alternative plan would be to engage the National Science and Technology Council, which could balance the operational elements of NOAA and the IPO with the research elements of NASA. However, the potential scientific and programmatic consequences of any of these possible options have not been studied in detail. Although the key agencies are beginning to propose studies, the committee is concerned that the eventual solution will be based largely on programmatic and fiscal constraints, not on the needs identified for climate research and monitoring.

Maintaining the status quo will continue to impede progress in understanding the processes of climate change and variability. Without such understanding, researchers will be unable to deliver sound scientific information and advice to policymakers to support informed decision making. Over the past 20 years, the power of satellite remote sensing to illuminate global-scale processes, especially in the area of short-term predictions, has become clear.

---

[1]However, there is growing evidence that long-term trends associated with climate will have significant economic impacts. For example, if the frequency of hurricanes on the U.S. East Coast changes, then the property insurance industry will have to adjust its investment and pricing policies. Similarly, long-term changes in precipitation in the Pacific Northwest may force the timber industry to alter its cutting cycle.

However, progress toward a climate observing system remains problematic. The IPO has recognized the requirements for climate observations through its inclusion of stability requirements for certain variables. The longer design life for the NPOESS sensors is also helpful. Nevertheless, the IPO has no mandate for climate observation. NASA/ESE has recognized the need to coordinate more closely with operational programs such as NPOESS to ensure long-term systematic measurements for critical climate variables. Activities such as the NPOESS Preparatory Project are tangible examples of such cooperation. However, from the perspective of the climate research community, much remains to be done. The overall integration of both operational and research components needs to be actively managed so that there is an effective process to achieve regular rebalancing in light of technological and scientific advances.

- NPOESS offers a unique opportunity to establish a satellite-based climate observing system. Although the NPOESS and NASA EOS missions as currently planned may not be optimum for climate research, many of the critical components are already in place. These include an initial commitment to data stability on the part of the NPOESS IPO, an active program of data analysis and data product validation by NASA's Earth Science Enterprise, and an active plan for NASA and NOAA collaborative missions such as the NPOESS Preparatory Project. The committee is concerned, however, that budget pressures, shifting programmatic interests, and lack of overall vision and leadership may continue to inhibit the establishment of a coherent Earth observing system for climate research and monitoring.

RECOMMENDATIONS

The following recommendations derive from consideration of the common issues associated with the space-based measurement of climate variables and committee concerns related to the conduct of climate research. They are directed to the climate research community, the NASA Earth Science Enterprise, and the NPOESS Integrated Program Office.

Recommendation 1.

Climate research and monitoring capabilities should be balanced with the requirements for operational weather observation and forecasting within an overall U.S. strategy for future satellite observing systems.

NPOESS represents a significant step forward in the nation's ability to produce and deliver short-term observations and predictions on Earth processes for protection of life and property. The present U.S. remote sensing strategy also contains many of the critical elements necessary for climate research and monitoring. Executive level initiative is essential to reconcile these requirements and to ensure that the nation's needs for short-term predictions are balanced with the needs to deliver scientifically based information on longer-term climate processes. Although the NPOESS platforms represent an essential component for stable, long-term observations, the present approach does not ensure that climate research needs will be met. Coordination of the needs for systematic and process measurements, sustainable and innovative technology, satellite and ground-based observations, and rapid delivery of data as well as ongoing, creation of long-term, consistent data sets relies primarily on a set of ad hoc agreements. A rigorous plan is needed, within the constraints of resources and assets, for achieving a rational balance between short-term (weather) and longer-term (climate) objectives. This balance should be evaluated on a periodic basis as scientific, programmatic, and fiscal constraints change.

The committee proposes the following specific actions to achieve this recommendation:

- **The Executive Branch should establish a panel within the federal government that will assess the U.S. remote sensing programs and their ability to meet the science and policy needs for climate research and the requirements for operational weather observation and forecasting.**

Such a panel would monitor agency efforts to provide balanced consideration of operational and climate research requirements. Elements to be considered include the research missions undertaken by NASA/ESE, the operational missions of IPO and NOAA, as well as ground-based networks and associated data systems. Remote sensing missions to be considered include polar, low inclination, and geostationary orbits. The panel should meet on a regular basis to assess the evolution of climate research and technological capabilities and to recommend adjustments to the observing system where appropriate.

- **The panel should be convened under the auspices of the National Science and Technology Council and draw upon input from agency representatives, climate researchers, and operational users.**

The panel would maintain cognizance of all the critical elements of the observing system that are relevant to operational uses and to climate research. It would pay close attention to the overall U.S. strategy to meet both sets of requirements, within the constraints imposed by schedule, funding, technology, and programmatics. It would recognize that there is not an "optimum" system but ensure that climate research requirements are considered along with other requirements.

- **The panel should convene a series of open workshops with broad participation by the remote sensing and climate research communities, and by operational users, to begin the development of a national climate observing strategy that would leverage existing satellite-based and ground-based components.**

The workshop process would establish the science priorities for climate observations and contribute to the development of a strategy that best meets these priorities within fiscal, schedule, technical, and programmatic constraints. Particular attention should be paid to the emerging plans of NASA/ESE, leveraging these with the capabilities of IPO/NPOESS programs. Although the initial focus would be on NASA and IPO, attention should also be paid to the other national and international components of the observing system as well as to modeling, data analysis, and algorithm development.

Recommendation 2.
The Integrated Program Office for NPOESS should give increased consideration to the use of NPOESS for climate research and monitoring.

The development of NPOESS should greatly improve short-term observations and predictions. Because of its emerging commitment to data stability, long-life platforms, and more capable sensors, there is an opportunity to develop NPOESS as part of the U.S. climate observing strategy without jeopardizing its focus on operational short-term forecasts.

The committee proposes the following specific actions to achieve this recommendation.

- **The IPO should consider the climate research and monitoring capabilities of NPOESS along with other NPOESS requirements.**

The proposed panel on satellite-based measurements for climate research should review and provide advice to the IPO as its plans for NPOESS mature, including its overall plans for sensor characterization, algorithm development, product validation, data system design, and so on. However, this approach should not be construed to mean that NPOESS should become a climate research mission, but rather that climate research requirements will be considered along with other requirements.

- **For those NPOESS measurements that are deemed to be critical for climate research and monitoring, the IPO should establish a science oversight team with specific responsibilities for each associated sensor suite.**

Because climate research requires the close involvement of scientists in all aspects of sensor development and operations, the IPO should establish mechanisms that go beyond the traditional contractor-selected teams. Climate science teams should be selected through a competitive process, perhaps in coordination with NASA. These science teams would function much like a NASA science team, developing and documenting algorithms, validating data products, and so on, but would focus on climate research.

- **The IPO should begin to establish plans for sensor calibration and data product validation as well as for data processing and delivery that consider the needs for climate research.**

Although the NPOESS sensors will not be launched for several years, climate research and monitoring will impose unique requirements on data systems and data quality. By utilizing the capabilities of NASA (where feasible), it may be possible to meet these requirements at a reduced cost to the IPO. Since NPOESS sensor development will begin soon, such efforts should start now to ensure that proper sensor tests are designed and implemented and critical information on sensor performance is maintained.

Recommendation 3.
The NASA Earth Science Enterprise should continue to play an active role in the acquisition and analysis of systematic measurements for climate research as well as in the provision of new technology for NPOESS.

Through its funding of climate research and its investments in technology, NASA can significantly enhance the climate research and monitoring capabilities of NPOESS and other operational missions. Activities could include the flight of complementary sensors and support for sensor characterization, as well as sensor calibration and data product validation, research and analysis, data reprocessing, and development of new technology.

The committee proposes the following specific actions to achieve this recommendation:

- **NASA/ESE should develop specific technology programs aimed at the development of sustainable instrumentation for NPOESS.**

Operational missions have stringent cost (for both development and operations) and longevity requirements, whereas the development of research instrumentation often faces fewer such constraints (especially in the short-term technology-driven missions in NASA/ESE's Earth Probes program). This discrepancy inhibits commitment by an operational agency to incorporate new technology into its programs. NASA should allocate some of its technology investments specifically to incentives designed to encourage the transition of technology to operational status, such as lowering the costs and improving the reliability and performance of candidate sensors. This might also include innovative methods for ensuring data continuity and sensor calibration. Such an approach can work effectively only if NASA's ESE and the NPOESS IPO cooperate at all stages of the process.

- **NASA/ESE should ensure that systematic measurements that are integrated into operational systems continue to meet science requirements.**

Although NASA/ESE intends to shift long-term responsibility for systematic measurements to operational programs, this may require that NASA/ESE continue to allocate resources to these measurements. Operational measurements may need enhanced capability, sensors may require more extensive calibration, and data will need to be analyzed and reprocessed. Operational programs may provide stable orbiting platforms, but not necessarily all of the infrastructure required to accomplish climate research and monitoring.

- **NASA/ESE should continue satellite missions for many measurements that are critical for climate research and monitoring.**

Because of conflicting requirements, some of the measurements required for climate research may not be incorporated into operational programs such as NPOESS. For example, some may not be deemed critical for short-term forecasts, or they may simply be too costly. NASA/ESE should continue such critical measurements, while pursuing new technologies or approaches that may reduce the costs of such observations. It is unrealistic to expect that all of NASA/ESE's systematic measurement requirements can be met by NPOESS.

Recommendation 4.
Joint research and operational opportunities such as the NPOESS Preparatory Project (NPP) should become a permanent part of the U.S. Earth observing remote sensing strategy.

The Operational Satellite Improvement Program (OSIP) was a joint activity between NASA and NOAA to improve short-term weather forecasting. A similar program in the area of climate research could help to fill the serious gap that now exists between the NASA/ESE research missions and the operational missions such as NPOESS. Operational missions tend to be conservative, valuing highly reliable and proven sensor designs as well as sensors that are inexpensive to operate. Research missions often value cutting-edge technology and favor new designs and new scientific approaches. A joint program could buffer the rapidly changing mix of science and technology emerging from NASA with the more constant requirements and capabilities of operational programs.

The committee proposes the following specific actions to achieve this recommendation:

- **The NPP concept should be made a permanent part of the U.S. climate observing strategy as a joint NASA-IPO activity.**

A permanent R&D testbed evolved from NPP would provide a unique opportunity to infuse research requirements and innovative technology into NPOESS without disrupting the operational requirements. In much the way that Nimbus provided a regular series of satellite platforms, a permanent series of NPP-style platforms could fly both operational and candidate operational sensors that measure identical variables as well as test new sensor concepts. A permanent NPP would ease the integration of research and operational missions by reducing risk and increasing confidence.

- **Some space should be reserved on the NPOESS platforms for research sensors and technology demonstrations as well as to provide adequate data downlink and ground segment capability.**

Current NPOESS plans include limited spare satellite resources (space, weight, power, etc.) that could support flight demonstrations of new sensor concepts. These resources are needed to ensure continued evolution of NPOESS capabilities over the decades that it will be in operation. Such an approach is only prudent and will ensure that NPOESS continues its state-of-the-art capabilities in short-term forecasting as well as improving its capabilities for climate research.

- **NPP and NPOESS resources should be developed and allocated with the full participation of the Earth science community.**

Because of time limitations, current plans for NPP were developed without much input from the science community. If the NPP concept is to be a permanent program element, then it is essential that future implementations be developed with much broader participation. Long-term goals should also be established. Some of the spare resources on the NPOESS platforms should be allocated based on a competitive process to ensure that the best science is accomplished. Such a program could be administered jointly by NASA and IPO as joint research announcements.

REFERENCES

National Research Council (NRC). 1995. Earth Observations from Space: History, Promise, and Reality. Washington, D.C.: National Academy Press.

National Research Council (NRC). 1998a. Overview, Global Environmental Change: Research Pathways for the Next Decade. Washington, D.C.: National Academy Press.

National Research Council (NRC). 1998b. The Atmospheric Sciences: Entering the Twenty-First Century. Washington, D.C.: National Academy Press

National Research Council (NRC). 1999a. The Adequacy of Climate Observing Systems. Washington, D.C.: National Academy Press.

National Research Council (NRC), Space Studies Board. 1999b. "Assessment of NASA's Plans for Post-2002 Earth Observing Missions," short report to Dr. Ghassem Asrar, NASA's Associate Administrator for Earth Science, April 8.

Appendixes

A

Statement of Task

INTEGRATION OF RESEARCH AND OPERATIONAL SATELLITE SYSTEMS

Background NASA officials have long envisioned developing operational versions of some of the advanced climate and weather monitoring instruments planned for the Earth Observing System (EOS) afternoon (PM) satellite. In the 1995 EOS "Reshape" exercise, NASA adopted the assumption that some of the measurements in the second PM series would be supplied by the Department of Commerce (NOAA) and Department of Defense (Air Force) National Polar-orbiting Operational Environmental Satellite System (NPOESS). NASA is about to begin intensive planning for the EOS-PM mission. NASA is also examining the potential for advanced instruments on future versions of the NOAA GOES (Geostationary Operational Environmental Satellite) satellites to be integrated into the EOS program.

Integrating NOAA-DOD operational weather satellites into NASA's Earth Observing System program poses numerous interrelated technical and organizational challenges. By definition, the "operational" weather programs of NOAA and DOD must meet the needs of users who require unbroken data streams. Historically, development of operational instrumentation has been successful when managed with a disciplined, conservative approach towards the introduction of new technology. In addition to minimizing technical risk, minimizing cost has been an important factor in the success of operational programs, especially for NOAA.

Achieving NASA research aims on a satellite designed to meet the operational needs of the civil and defense communities will require agreement on joint agency requirements, and coordination of instrument development activities, launch schedules, and precursor flight activities. The proposed study will include an analysis of these issues, especially those related to (1) sensor design and development, (2) program synchronization, and (3) data continuity and interoperability.

Plan The proposed study will analyze generic issues related to the transition of NASA research satellite instrumentation for NOAA operational use. The study will focus in particular on observational priorities and technical issues related to the potential integration of the NOAA-DOD NPOESS satellite with the NASA EOS "PM" series of satellites. Among the key questions to be addressed are:

1. Sensors and Measurements
- How well do current NPOESS IORD requirements match NASA research requirements for the EOS PM-2 satellite series? Are there any overlaps with AM-2 or CHEM requirements?
- If additional capability is needed for climate monitoring goals, what is this capability and what are technical and programmatic implications?
- Are there instruments that could be added to the operational suite, e.g., a scatterometer or SAR? What issues must be addressed in adding capabilities of this kind?
- What are the requirements for on-orbit or ready-to-launch replacement instrumentation for research and operational goals? Are there common spares strategies that could serve both research and operational needs satisfactorily?
- What issues might arise should NPOESS be tasked to undertake new missions such as long-term climate monitoring?

2. Program Synchronization
- What are the critical milestones in integrating research and operational space systems? Are any disjoints apparent?
- What are possible approaches to establishing program flexibility to ensure that both research and operational missions are achieved in the face of inevitable schedule changes?

3. Data Continuity and Interoperability
- What are the highest priorities for continuous/interoperable research datasets?
- What are technical approaches to ensuring data (a) interoperability between research and operational sensors and (b) continuity in the face of evolving sensor technology?
- What is the status of data storage, retrieval, and access planning for research use of NOAA operational data or possible NPOESS-obtained climate data?

A report summarizing the findings and recommendations that address technical items (1) and (2) ('Sensors and Measurements') and 'Program Synchronization') will be the Phase 1 report. Item (3) 'Data Continuity and Interoperability' will be addressed in the Phase 2 report.

B

Acronyms and Abbreviations

4DDA	four-dimensional data assimilations
AATSR	Advanced Along Track Scanning Radiometer
ACE 1, 2, and 3	Aerosol Characterization Experiments
ADEOS-2 or ADEOS-3	Advanced Earth Observation Satellite, also, Advanced Earth Observing System
ADM	Angular Directional Model
AERI	Atmospheric Emitted Radiation Interferometer
AERONET	Aerosol Robotic Network
AIRMISR	Airborne Multi-angle Imaging Spectroradiometer
AIRS	Atmospheric Infrared Sounder
AirSAR	Airborne Synthetic Aperture Radar
ALI	Advanced Land Imager
AM-1 or AM-2	morning platform 1 or 2
AMSU-A	Advanced Microwave Sounding Unit A
AMSU-B	Advanced Microwave Sounding Unit B
AOS	Aerosol Observation System
ARM	Atmospheric Radiation Measurement
ASTER	Advanced Spaceborne Thermal Emission and Reflection Radiometer
ATMS	Advanced Technology Microwave Sounder
ATSR	Along Track Scanning Radiometer
AVHRR	Advanced Very High Resolution Radiometer
AVIRIS	Airborne Visible and Infrared Imaging Spectrometer
BOREAS	Boreal Ecosystem-Atmosphere Study
BRDF	bidirectional reflectance distribution function
CAC	Climate Analysis Center
C-band	microwave wavelength used in radar signals
CCD	charge-coupled device

CCN	cloud condensation nuclei
CDOM	colored dissolved organic matter
CEOS	Committee on Earth Observation Satellites
CERES	Clouds and the Earth's Radiant Energy System
CES	Committee on Earth Studies
CHEM	satellite to measure atmospheric chemistry
CMIS	Conical Scanning Microwave Imager/Sounder
CNES	Centre National d'Etudes Spatiales
CrIS	Cross-Track Infrared Sounder
CRWP	Climate Requirements Workshop Report
CSA	Canadian Space Agency
CZCS	Coastal Zone Color Scanner
DEM	Digital Elevation Model
DIAL	Differential Absorption Laser
DMS	dimethyl sulfide
DMSP	Defense Meteorological Satellite Program
DOD	Department of Defense
DOE	Department of Energy
EC	elemental carbon
ECS	EOSDIS Core System
EDR	environmental data record
ENSO	El Niño/Southern Oscillation
ENVISAT	Environmental Satellite (European Space Agency's first research polar platform)
EO-1	Earth Observer 1, first satellite of the New Millennium Program
EOS	Earth Observing System
EOS AM	Earth Observing System Morning Satellite
EOS PM	Earth Observing System Afternoon Satellite
EOSDIS	Earth Observing System Data and Information System
EOSP	Earth Observing Scanning Polarimeter
EP	Earth Probes
ERB	Earth Radiation Budget
ERBE	Earth Radiation Budget Experiment
ERBS	Earth Radiation Budget Satellite
EROS	Earth Resources Observation Satellite or System
ER-2	NASA-ER2 high-altitude research aircraft
ERS	Earth Resources Satellite
ESA	European Space Agency
ESE	Earth Science Enterprise
ESS	Earth System Science
ESSIPs	Earth System Science Information Partnerships
ESSP	Earth System Science Pathfinder
ESTAR	Electronically Scanned Thinned Array Radiometer
ETM+	Enhanced Thematic Mapper
EUMETSAT	European Organization for the Exploitation of Meteorological Satellites
FAO	Food and Agriculture Organization
FIFE	First ISLSCP Field Experiment
FMOC	Fleet Meteorology and Oceanography Center

FOV	field of view
FTS	Fourier transform spectrometer
FWHM	full width at half maximum (bandwidth)
GAC	Global Area Coverage
GAIM	Global Analysis, Interpretation, and Modeling
GCIP	GEWEX Continental-scale International Project
GCM	general circulation model
GCOS	Global Climate Observing System
GCRP	Global Change Research Program
GCS	gas-correlation spectrometer
GEO	geosynchronous Earth orbit
GERB	Geostationary Earth Radiation Budget
GEWEX	Global Energy and Water Cycle Experiment
GEWEX PILPS	GEWEX Project for Intercomparison of Land-Surface Parameterization Schemes
GHG	greenhouse gas
GLI	Global Imager
GLIS	Global Land Information System
GODAE	Global Ocean Data Assimilation Experiment
GOES	Geostationary Operational Environmental Satellite
GOFC	Global Observation of Forest Cover
GOME	Global Ozone Monitoring Experiment
GPP	gross primary productivity
GPS	Global Positioning System
GSFC	Goddard Space Flight Center (NASA)
GTOS	Global Terrestrial Observing System
HALOE	Halogen Occulation Experiment
HAPEX	Hydrological-Atmospheric Pilot Experiment
HDF	hierarchical data format
hh-polarization	horizontally polarized emitting and receiving radar signals
HIRDLS	High-Resolution Dynamics Limb Sounder
HIRS	High-Resolution Infrared Sounder
HIS	High-Resolution Interferometer Sounder
IAM	integrated assessment model
IASI	Infrared Atmospheric Sounding Interferometer
IGAC	International Global Atmospheric Chemistry Program
IGBP	International Geosphere-Biosphere Program
IGBP-DIS	IGBP data and information system
IGBP/IHDP LUCC	IGBP/International Human Dimensions Programme on Global Environmental Change/Land-Use and Land-Cover Change
IGOS	International Global Observing Strategy
IGPO	International GEWEX Project Office
IHDP	International Human Dimensions Programme on Global Environmental Change
IIP	Instrument Incubator Program
ILAS	Improved Limb Atmospheric Spectrometer
IOCCG	International Ocean Color Coordinating Group
IOM	Institute of Medicine
IORD	Integrated Operational Requirements Document

IORD-1	Integrated Operational Requirements Document (First Version)
IPCC	Intergovernmental Panel on Climate Change
IPO	Integrated Program Office
IR	infrared
IRS	Indian Remote Sensing (series of satellites)
ISLSCP	International Satellite Land Surface Climatology Project
JERS	Japan's Earth Resources Satellite
JPL	Jet Propulsion Laboratory
KVR-1000	the Russian panchromatic band
LAI/FPAR or LAI/*f*PAR	leaf area index/fraction of absorbed photosynthetically active radiation
Landsat	Land Remote Sensing Satellite
LBA	Large Scale Biosphere Atmosphere Experiment in the Amazon
L-band	low-frequency
LEO	low Earth orbit
LightSAR	Light Synthetic Aperture Radar
LITE	Lidar In-Space Technology Experiment
M-AERI	Marine-Atmospheric Emitted Radiation Interferometer
MAS	MODIS Airborne Simulator
MASTER	multiband optical and thermal infrared sensor
MCSST	Multi-Channel SST (algorithm)
MERIS	Medium-Resolution Imaging Spectrometer
METEOR	Russian Operational Weather Satellite
METEOSAT	European Geostationary Meteorological Satellite
METOP	Meteorological Operational satellite
METOP-3	EUMETSAT Operational Polar Orbiter 3
MIPAS	Michelson Interferometer for Passive Atmospheric Sounding
MISR	Multi-angle Imaging Spectroradiometer
MLS	Microwave Limb Sounder
MODIS	Moderate-resolution Imaging Spectroradiometer
MSG	METEOSAT Second Generation
MSS	Multispectral Scanner
MSU	Microwave Sounding Unit
MWIR	medium wavelength infrared
NASA	National Aeronautics and Space Administration
NASDA	National Space Development Agency-Japan
NAVO	Naval Oceanographic Office
NCEP	National Centers for Environmental Prediction/Climate Prediction Center
NDVI	Normalized Difference Vegetation Index
NEΔL	noise-equivalent delta radiance
NEMO	Naval Earth Map Observer
NESDIS	National Environmental Satellite, Data, and Information Service
NetCDF	network Common Data Format
NEP	net ecosystem productivity
Nimbus-7	NASA Environmental Research Satellite Series, 7

NIR	near infrared
NIST	National Institute of Standards and Technology
NLSST	Non-Linear SST (algorithm)
NMP	New Millennium Program
NOAA	National Oceanic and Atmospheric Administration
NPOESS	National Polar-orbiting Operational Environmental Satellite System
NPP	NPOESS Preparatory Project
NRC	National Research Council
NWP	Numerical Weather Prediction
OCTS	Ocean Color and Temperature Scanner (on ADEOS-1)
OMI	Ozone Monitoring Instrument
OMPS	Ozone Mapping and Profiler Suite
OPT	Ozone Processing Team
ORACLE	Ozone Research with Advanced Cooperative Lidar Experiments
Orbimage-2	Orbital Science's Imaging Satellite 2
OSIP	Operational Satellite Improvement Program
OTIS	Ocean Thermal Optimum Interpolation System
PALSAR	phased-array type L-band synthetic aperture radar
P-band	microwave wavelength used in radar signals
PI	principal investigator
PICASSO-CENA	Pathfinder Instruments for Clouds and Aerosols using Spaceborne Observations-Climatologie Etendue des Nuages et des Aerosols
PM-1 or PM-2	afternoon platform 1 or 2
PMC	polar mesospheric cloud
POAM	Polar Ozone and Aerosol Measurement
POES	Polar-orbiting Operational Environmental Satellite
POLDER	Polarization and Directionality of the Earth's Reflectances
PSC	polar stratospheric cloud
RADARSAT	Canada's Synthetic Aperture Radar Satellite
RAL/UK	Rutherford Appleton Laboratory/United Kingdom
rms	root mean square
SAFARI	Southern African Fires Atmosphere Research Initiative
SAGE	Satellite Aerosol and Gas Experiment
SAM	Stratospheric Aerosol Measurement
SAR	synthetic aperture radar
SBUV	Solar Backscatter Ultraviolet
SCIAMACHY	Scanning Imaging Absorption Spectrometer for Atmospheric Cartography
SeaWiFS	Sea Viewing Wide-Field-of-View Sensor
S-GLI	S-Global Imager
SIMBIOS	Sensor Intercomparison and Merger for Biological and Interdisciplinary Oceanic Studies
SIR-C	Shuttle Imaging Radar-C
SIR-C/X SAR	Shuttle Imaging Radar-C/X-band synthetic aperture radar
SMMR	Scanning Multichannel Microwave Radiometer
SNR	signal-to-noise ratio
SOLVE	SAGE III Ozone Loss and Validation Experiment
SPOT	Systeme pour l'Observation de la Terre (France)

SSBUV	Space Shuttle Solar Backscatter Ultraviolet
SSMI	Special Sensor Microwave/Imager
SSM/T/1	DMSP Microwave Temperature Sounder
SSM/T/2	DMSP Microwave Water Vapor Profiler
SST	sea surface temperature
SVAT	soil vegetation atmosphere transfer
TARFOX	Tropospheric Aerosol Radiation Forcing Observation Experiment
TDR	time domain reflectivity
TES	Tropospheric Emission Spectrometer
TIROS	Television Infrared Observation Satellite
TIROS-N	fourth-generation NOAA polar-orbiting operational environmental satellite
TM	Thematic Mapper
TOA	top of the atmosphere
TOMS	Total Ozone Mapping Spectrometer
TOPC	Terrestrial Observation Panel for Climate
TOVS	TIROS Operational Vertical Sounder
TRMM	Tropical Rainfall Measuring Mission
UARS	Upper Atmosphere Research Satellite
USDA	U.S. Department of Agriculture
USGCRP	U.S. Global Change Research Program
UV	ultraviolet
VCL	Vegetation Canopy Lidar
VIIRS	Visible/Infrared Imager and Radiometer Suite
VNIR	visible and near infrared
VOCs	volatile organic carbons
vv-polarization	Vertically polarized emitting and receiving radar signals
WCRP	World Climate Research Program
WMO	World Meteorological Organization
WOCE	World Ocean Circulation Project
x-band	Microwave region of the electromagnetic spectrum defined by wavelengths of approximately 2.4 to 3.8 cm